The Origin of Feces

What Excrement Tells Us About Evolution, Ecology, and a Sustainable Society

David Waltner-Toews

ECW Press

Published by ECW Press
2120 Queen Street East, Suite 200, Toronto, Ontario, Canada M4E 1E2
416-694-3348 / info@ecwpress.com

LIBRARY AND ARCHIVES CANADA CATALOGUING IN PUBLICATION

Waltner-Toews, David, 1948–
The origin of feces : what excrement tells us about evolution, ecology, and a sustainable society / David Waltner-Toews.

Includes bibliographical references.

ISBN 978-1-77041-116-6
ALSO ISSUED AS: 978-1-77090-396-8 (PDF); 978-1-77090-397-5 (EPUB)

1. Feces. I. Title.

QP159.W35 2013 573.4'9 C2012-907524-8

Editor for the press: Crissy Boylan
Cover design: David Gee
Cover image: ©Antagain/iStockphoto.com
Printing: Trigraphik | LBF 5 4 3 2 1

The publication of *The Origin of Feces* has been generously supported by the Canada Council for the Arts which last year invested $20.1 million in writing and publishing throughout Canada, and by the Ontario Arts Council, an agency of the Government of Ontario. We also acknowledge the financial support of the Government of Canada through the Canada Book Fund for our publishing activities, and the contribution of the Government of Ontario through the Ontario Book Publishing Tax Credit. The marketing of this book was made possible with the support of the Ontario Media Development Corporation.

Canada Council for the Arts / Conseil des Arts du Canada
Canada
ONTARIO ARTS COUNCIL / CONSEIL DES ARTS DE L'ONTARIO
Ontario

PRINTED AND BOUND IN CANADA

"... all prejudices come
from the intestines."

— FRIEDRICH NIETZSCHE

This book is for the late Bruce Hunter
(August 3, 1950–October 19, 2011),
friend, colleague, farmer, veterinarian,
ecohealth pioneer, global citizen,
a man who really knew his shit.

I miss him.

PREFACE

DUNG BEETLES AND THE GIRL ON THE AIRPLANE

Our single-prop four-seater plane made a low swoop over the short runway, sending the gangly giraffes, monkeys, and impala loping and leaping over the red dirt into the shelter of the trees in the wooded grassland. In the days that followed, from our safari truck, we saw lion cubs and lionesses playing, kitten-like, in the shadow of our vehicle. We were charged by a huge female elephant, her ears fanned out wide to the sides. "She's bluffing. Sit still," said the guide. We froze. She stopped suddenly, twenty meters away, turned, and walked slowly back to rejoin her family. We thawed in the slow heat. Over gin and tonic, we watched the hippos lolling in the river; we heard them grunting and crashing past our cabin at night. We glimpsed the hyenas retreating into the darkness as our Maasai guide, lantern and spear in hand, led us back to our cabin after dinner. Bee eaters — bright flashes of green and white and black — darted and swooped into and out of small caves in the earthen cliffs along the river as our boat cruised past. Storks, stilts, spoonbills, and sandpipers trod daintily in the shallows.

Several days into our safari retreat, in the baking late-morning heat, ten of us followed two guides through green-streaked, straw-colored grasses, between ash-gray tree skeletons, flat-topped *Terminalia spinosas* (those classic African trees that photograph so well against the setting sun), and gnarled, deep green shrubs. It was the dry season in the Selous Game Reserve in Tanzania, the most astounding and exuberantly diverse wildlife park in all of Africa. Almost everyone else in our group was on the lookout for more elephants, lions, Cape buffalo, and warthogs, and nervously excited about the possibility of encountering them without the benefit of an escape vehicle. The effect was enhanced by the fact that Mpato, one of the guides, was carrying a rifle big enough to bring down a large, angry beast, should that be necessary.

To the dismay of the others on our walkabout, I had also corralled the guides into helping me find animal scat. What kind of an idiot comes to Africa to look at shit? Well, I'm a veterinarian, so maybe it fits.

A small white pile of feces, away from a lake-sized widening in the Rufiji River, was probably hyena. The white indicated a carnivore, the crushed bones of its prey bleaching in the hot sun. A similar whitening (and hence carnivore) pile closer to the lake was probably from a crocodile. Small pellets in a heap were male impala marking territory, as well as providing nitrogen and phosphorus for plants. Female impala scat were a more scattered whisper across the landscape, and less a megaphone to announce ownership. Buffalo dung pies were flat, circular pats like domesticated cow dung, but firmer. Zebra feces, as we might have expected, were more horse-like, kidney-shaped, and darker than buffalo dung. Hippo dung on the path was like that of an elephant — a big cylindrical sausage — but wetter and with less fiber. The increased water content is associated with a lower concentration of nutrients, meaning it is less attractive as a food source for other species than elephant dung. The lower fiber concentration also means hippo dung is less useful for making paper than elephant dung. Hippos, who cannot see well, use a tactic that would have saved Hansel and Gretel, with their dainty bread crumbs, considerable grief. They mark the path to and from the river with their dung when they go on nighttime food foraging and pillaging expeditions, so they can use it to find their way back from the riverbank. Hippos are thus moving nutrients from the water to the land and, when they defecate in the water (which they also do), back to aquatic systems.

Each of the scats I saw along the trail in Tanzania told me something about the animals and their ecological places in this landscape, and the ways in which they have physically changed the landscape (eating plants, stomping paths, transferring seeds). Nevertheless, looking at animal scat was still vaguely unsatisfying. What I really wanted to see, much to the dismay of my fellow safari-ists, most of whom were honeymooners, were the queens of the wilderness — dung beetles. I think I was, at first, somewhat amused by the notion that there are animals that literally eat shit. What did they look like? How did they do it?

I was heading back to my tent after the walk when Eduard, one of our two guides, came running after me. He and Mpato had found some dung beetles near the camp, hard at work! Strangely, as Eduard told me about this amazing find, the other tourists disappeared.

"You go," said my wife. "I have some important reading to do."

"Their loss," I thought as I followed Eduard a hundred

meters down the path to where Mpato was guarding a large pile of elephant dung with his hunting rifle. I crouched next to him, the sun beating down on my head, the sweat trickling down the middle of my back.

If I was disappointed with what I saw, I did not show it, although I had secretly wished for a re-creation of the scene described by two researchers in 1974; they had observed 16,000 dung beetles attack one 1.5-kilogram pile of elephant dung and eat, bury, and roll away the whole pile within two hours. I had to remind myself that seeing even two thumb-sized black beetles scuttling around the dung-pile was more than I should have expected this time of year. In the dry season, most dung beetles in the area were already hunkered underground in front of their television sets, watching reruns of *Bushland Idle*.

One of the two beetles was industriously pulling down chunks from the heap of dung and tamping them down to form a ball. Nearby, another beetle was busy burying his ball. The sandy earth shifted up and down with his subterranean digging movements. We squatted in the hot sun, watching the manic, obsessive beetles at work. If Mpato and Eduard were puzzled at my interest, they were too polite to show it.

Just when we thought he seemed done, when the walnut-sized sphere of dung — twice his size — looked perfect to me, the beetle would pull down another piece and tap it into some imperceptible imperfection. Finally he started rolling the ball away from the dung pile, uphill, over twigs, his hind end up against the ball, front end down on the ground. He strained and tumbled over a small branch and rolled into a little gully, still clinging to his large ball. Then he clambered up, had a look around at the lay of the land, and started pushing again. Periodically, he dug down into the earth or under the leaf litter, then came back up and resumed pushing. He pushed and tumbled his way some eight meters from the dung pile, uphill almost all the way, until he finally started digging seriously, his ball sinking into the leaf litter, heaving once or twice, and then sinking from view.

The other beetle, nearest to the dung pile, made another, smaller, ball, dug a hole, pushed in the ball, and then came back out and pulled in more dung. Both beetles were probably males, who are stronger than females, and who usually do the rolling. Sometimes, males and females will roll the ball together, bury it, dig a tunnel nearby, have sex, and then insert the fertilized eggs into the ball. If another male comes along that wants sex with the female, the two males lock horns and try to push each other out of the tunnel. According to some researchers, these battles for sex have resulted in some beetles that can pull more than 1,000 times their own weight — a person, say, pulling six double-decker buses full of people. It would not be the first time that a male has gone to the gym to improve his chances of finding a sexual partner.

In the case of the beetles I was watching, I could see no females around. Perhaps they had already bred. Or perhaps they were single males, unlucky in love, tucking away a bit of nourishment for the rest of the dry season, in which case there was some poignancy to the scene I was observing.

Whatever their personal stories, it occurred to me that what I was watching was surely more than a curiosity. A huge amount of energy was being expended by these animals to build and bury these large nutri-balls of manure. While this energy use makes sense for them as individuals in that it enables the survival of their young, it also makes sense in much larger terms. For what nourishes their young also nourishes the landscape, which provides food for the elephants, who provide the shit that allow the dung beetle babies — as well as many other species — to live through the dry season. A 2008 review of the ecological literature on dung beetles by the Dung Beetle Ecology Working Group of ScarabNet summarized contributions of various dung beetles to cycling of nutrients, enhancing plant growth, controlling parasites, and dispersing seeds. The unpaid and unsung work of these beetles in parasite control, pasture improvement, and reduction in greenhouse gases has contributed hundreds of millions of dollars to global agriculture.

There are dung beetles on every continent except Antarctica. Many are adapted to work not just in particular landscapes, but are also fastidious about the kinds of dung they eat. The dung beetles of Australia, for instance, accustomed to marsupial dung, wanted nothing to do with cow pats. For almost 200 years after European settlement, as the geographic extent and number of cattle increased, the problems associated with dung flies and general stinky unpleasantness increased. In the 1970s, in an experiment considerably more successful than the ill-advised importation of cane toads to kill cane pests, some twenty species of African dung beetle were selected and imported. These brought the cattle manure situation under control and helped improve soil and pasture quality. Unfortunately, the widespread use of drugs to treat insect pests on cattle, a process which also kills dung beetles, may now be reversing some of those gains.

Like all of us, dung beetles can be classified by their physical characteristics, their genetics, and their behavior. The ones I observed in Tanzania were rollers: they wrap a ball of feces around their eggs and then bury them. When the larvae hatch, they eat the dung around them, which must mean that the dung itself is rich in energy and nutrients — energy and nutrients that would otherwise leak away into the environment. Not all dung balls are as small and malleable as the ones I saw. The balls created by some Indian scarabs have been mistaken for old stone cannonballs, because the clay with which the beetles covered them had hardened. Tunnelers bury the feces, and dwellers, the laziest of the bunch, just live in it. Some species hang around watching a neighbor roll up a large ball, and then, when he skitters off to get more dung, they steal it!

While there are ongoing debates among coleopterists (people who study beetles and their habits) over details of beetle evolution, what is not in doubt is that excrement-eating beetles have been around for a long, long time. Early scarab beetles (*Scarabaeoidea*), from which modern dung beetles evolved, appear to have been around for more than 150 million years, and "true" dung beetles (the ones that actually eat dung), about 40 million years. Evidence from fossilized dung pieces, called coprolites, suggests that dung beetles developed commensal relationships with dinosaurs in the Mesozoic period, that is, more than 65 million years ago, when the continents were slowly creaking and groaning on their rebellious, separate ways, budding off from mother-father Pangaea and carrying with them a variety of species to start new lives elsewhere on the globe. If we are serious about staying around on the planet for a long time and maintaining a meaningful, if challenging, lifestyle, we could do worse than to consult the dung beetles.

Many scientists love to get close, really close, to nature,

to focus on a few small things to the exclusion of all else. So, if you search the literature on dung beetles, you will find references to anywhere from 5,000 to more than 7,000 species, divided into twelve tribes, with a couple hundred new species being described every year. There are about 100 species of dung beetle devoted to elephant dung alone. Part of the problem of keeping count is that the nomenclature has been unclear, with some writers slipping back and forth between "dung beetles" and "scarabs."

Although all scarabs are great recyclers, not all scarabs are dung beetles. The *Scarabaeoidea* are a superfamily, including within it families whose names include such luminaries as sand-loving, enigmatic, earth-boring, rain, and bumblebee beetles. They include fungivores (which eat fungi), herbivores (eat plants), necrophages (eat dead things), carnivores (eat other animals), saprophages (eat any kind of decaying organic matter), and — the ones that interest us in this book — the coprophages (eat shit). Species in the subfamily *Scarabaeinae* are sometimes referred to as the "true" dung beetles, since most of them feed almost exclusively on feces. The taxonomic boundaries of their families, subfamilies, and tribes may be unclear, but all can claim royal heritage. Scarabs were revered, and rightly so, by the ancient Egyptians — another case, and there are many, where traditional religious practices helped preserve important ecological functions.

To the ancient Egyptians, rather than representing filth and feces, the scarab suggested death and rebirth, renewal and resurrection. Like the god Khepri, who created himself from nothing, rolled the sun through the darkness, and, *voilà,* presented it new each morning, the scarab rolled its sphere to the underworld and, fifteen to eighteen weeks later, was reborn. Hence, the name and picture used to depict the scarab was "to come into existence," and hence the celebration of the scarab in precious metals and stones, bones and ivory, in funeral rites of the region and in "mummy" adventure B-movies.

While most tourists in Africa pay attention to the so-called charismatic megafauna such as impala and elephants, few are attentive to the much smaller creatures that help create the landscapes in which those larger animals live, and which enable them to survive. Early in human evolution, it made sense to pay attention to animals you could eat, or that might be a threat to you. In the twenty-first century, it is the loss of the animals we don't see, and for which we don't see an immediate use — dung beetles for instance — that may pose the biggest threat to us. Dung beetle safaris should be the next great opportunity for ecotourism.

As I watched the two East African beetles at work, I reflected that these animals were more than a curiosity. They embodied, in many ways, a question that had been eating away inside me, like dung beetle larvae in a dung ball, over decades of teaching epidemiology of foodborne and waterborne diseases. How and why has excrement — which is absolutely necessary for the resilient functioning of our planet, and which has, in fact, been a solution to a myriad of biological problems thrown up by the long haul of evolution — become, in the past mere few thousands of years, a problem to be solved? When was the challenge of dancing through this amazing web of life-giving-life reduced to an issue of sustainable manure management?

Every day, all around the world, by eating or burying what others consider waste, dung beetles turn water into wine, contaminated refuse into livable landscapes. They are the Rumpelstiltskins of the animal world, weaving gold from dung-straw. They close the feedback loops of nutrients and energy that are essential for the resilience and health of the ecosystems that are our home. Can we learn from them? Is it important that we learn from them? Should we give, as they say, a shit?

Over the past few years, variations of this story circulated on the internet:

> A man in a dark blue suit makes his way down the aisle of the airplane and is delighted to find that there is a pretty woman in the seat next to his. He takes off his suit jacket, carefully folds it, and places it in the overhead bin. Then he sits down, loosens his tie, puts his computer under the seat in front of him, and looks over at her. She is reading a book and does not look up. He clears his throat. "I always find that the flight goes faster if you strike up a conversation with your fellow passengers."
>
> The woman slowly closes her book, brushes her dark, wavy hair away from her eyes, and asks, "Okay, what would you like to discuss?"
>
> The guy says, "Oh, I don't know. Anything. How about nuclear power?"
>
> The woman sighs. "That could make for a pretty interesting conversation. But let me ask you a question first, okay?"
>
> "Sure," he says. "Why not?"
>
> "A horse, a cow, and a sheep all eat grass, but the sheep excretes pellets, the cow plops out patties, and the horse puts out something that looks like rye buns. Why is that?"
>
> The guy shrugs his shoulders and grins at her. "I don't know."

She stares at him. "So, do you really think you're qualified to discuss nuclear power when you don't know shit?"

She opens her book and resumes reading as the plane takes off.

It is always risky to deconstruct a joke, and even more risky if the joke involves some slightly impolite topic. But I will do so, because the story I am about to tell in this book is woven around the pun in the airplane joke. The unhappy male traveler is caught in the snare of not knowing much about excrement and is therefore accused of not knowing much about anything. In fact, the woman's use of the "s" word is a triple threat. She is not only being sarcastic about his knowledge of biology, and his lack of knowledge about life in general, but she is using a word that is generally reserved for the drunken carelessness of bars or hormone-driven bravado of locker rooms. Hence she is at once implying that the topic is profoundly important and maligning it. As a beautiful young woman well versed in the techniques of deflecting unwanted male interest, she can get away with this contradictory put-down, to which there can be no winning response.

I empathize with the man, in no small part because he is a man on an airplane trying to make diverting conversation, which is a situation in which I have often found myself. I also empathize with the hapless fellow because, as a veterinarian, I have found myself on occasion with my arm up a cow's butt, or examining feces from a dog, or lying in a gutter full of animal excrement, or even, more recently, seriously studying all manner of human and animal diseases transmitted by what is called the fecal-oral route (which you won't find on Google Maps or MapQuest). Nevertheless, I realized, somewhat late in my professional life, that I actually knew very little about the stuff.

Staring at those dung beetles, a mad white man under the blazing midday sun, I realized that before we can understand why they are so important, not just in themselves, but as an exemplar of the kinds of solutions we might devise to face the complex challenges of surviving through the twenty-first century, we must understand the larger context, the universe in which we are all indigenous peoples.

Through decades of work on human foodborne and waterborne diseases, I have discovered that excrement, and how we think about it, is profoundly linked to everything I care deeply about — culture, food, health, ecological sustainability. Especially ecological sustainability, without which nothing else exists. The lovely young lady on the plane is absolutely right. Really, if you don't know road apples from cow pies, shit from fertilizer, you probably shouldn't be talking about nuclear power; the fact that most people in the power business (political, economic, energy) cannot speak intelligently on the subject indicates a profound ignorance, rooted in a deep alienation from our most essential biological selves.

The way we treat excrement is a choice that feeds certain species and steals from others, destroys certain ecosystems and builds up others. Why has shit become such a public health and environmental problem, when it would seem to be such a smorgasbord of ecological opportunities? Unless we change how we think about shit, we are doomed to forever live in it. Or perhaps, more accurately, we already do live in it, and always will, and unless we change how we think about it, we will continue to create very unfortunate problems for ourselves.

This book is about rectifying (no pun intended) our

ignorance, and achieving that wisdom of discernment. But more than that, this book is about moving through discernment to the unified reality that underlies everything, the nameless ground of all being, to what the ancient Chinese called the Tao and the tribes of the Middle East, the "I Am." If this seems to you too ambitious a task for a small book about feces, if this seems too large a burden to place on such a small beast as a dung beetle, then, dear reader, you are exactly the person for whom this book is written.

Despite our journey from the inchoate soups of the early universe to the highest achievements of modern civilization, we are still, deep inside, animals. Whatever our station in life, rich or poor, powerful or downtrodden, priest or anarchist, man-god or ape-man, we still have to urge the passage of materials through our bowels. If we can understand this substance that emerges from us, from all animals, as an ecological unifying principle, which goes back to our evolutionary origins and the roots of our belonging, then we can, with serenity and happiness, deal with all the visible shit that surrounds us.

Read on. Know shit.

CHAPTER 1

WHAT FROM THE TONGUE FALLS

Fictional character Tante Tina Wiebe, in a dramatic monologue recounting family stories about Christmas, says that, whatever else has changed about Mennonite culture, the men still walk as if they are "bringing in the cows," but then laments that

> ". . . maybe too much sometimes
> what was once on the boots clinging
> now from the tongue falls . . ."[1]

Before one can tell a story, one must have words. With words come culture, and with culture come taboos and conflicts, things we don't talk about, and don't talk about not talking about, even if we are sitting in a pile of it. If we cannot name "what from the tongue falls," how can we possibly, seriously, address all the other dimensions of

1 Waltner-Toews, D. 1995. *The Impossible Uprooting*. Toronto: McClelland & Stewart. Page 85.

excrement? How can we unleash the incredible power of excrement if we don't know shit?

Shit is what sociologists and scientists call a wicked problem.

The social planners who introduced the idea of wicked problems in the 1970s differentiated them from what were considered to be the "tame" problems addressed by conventional science. Wicked problems, they asserted, are poorly bounded and contradictory. They are difficult to solve because information is incomplete, or the requirements of those who want the problem solved keep changing. They can be defined from a variety of apparently incompatible perspectives, so that there is neither a definitive problem formulation nor an optimal solution. Worst of all, the solutions to some aspects of the problem may create or reveal other problems.

Many public health and environmental researchers and managers are faced with such wicked problems. For instance, we can get rid of malaria by spraying insecticides and filling in wetlands, and stop the spread of some serious viruses by destroying farm animals and wildlife. The unintended, long-term consequences of these solutions for ecological sustainability, human health, and the livelihood of farmers, however, are worrisome and huge. Another example: let us say — as did Henry IV of seventeenth-century France and the American Republicans of the 1920s — that we aspire to get a "chicken in every pot" at least once a week, with the good intention of improving people's nutrition. By the 1960s, we discovered that we could do this by creating intensive farms housing millions of animals and by promoting free global trade. But this same strategy also enables the global spread of disease-causing bacteria like *Salmonella*, and puts millions of small farmers out of business. More data, better tests, more refined laboratory techniques, or more sophisticated modeling will do little to resolve such wicked problems.

What places shit among the wickedest of wicked problems is that, with all the ecological and public health impacts it carries, we don't even have a good, common language to talk about it. We can use precise technical terms when we want the engineers to devise a solution to a specific organic agricultural and urban waste problem (processing this agglomeration of shit and other waste into what they call biosolids, for instance). In so doing, however, we alienate the public, who are suspicious of words like biosolids. This public will need to pay for the filtration and treatment plants. They suspect that the solution to chicken shit in the water might not be a better filtration plant, but they don't have the language to imagine and discuss what the alternatives might be. Through the language we use, we separate the two things that absolutely need to be integrated: the popular political imagination and the scientific and technical understanding of the substantive issues.

To return to the "chicken-in-every-pot" challenge: we solve the problem of providing animal protein to large numbers of people through intensive livestock farming. Now we have a few very big farms where there used to be many smaller ones. As a consequence, where once the chicken shit was scattered widely across, and absorbed into, the rural landscape, we now have gigantic, localized piles of excrement. This localized overproduction of shit results in water contamination and public health problems, which are solved through "sustainable manure management" such as feeding the chicken shit to cows, or building giant bio-gas plants to generate electricity. So now, in this theoretical example, we have our cows dependent on chickens for nitrogen intake, and our electricity supply dependent on having large farms to produce lots of shit.

How, then, can we get a handle on this slippery, amoeba-like problem? In this book, I shall poke at it from several perspectives: excrement as a problem of language, as a public health problem, and as an ecological problem. These three perspectives are used by society to characterize excrement and have resulted in three different, and often conflicting, solutions to its perceived problems. Public health–oriented solutions can make ecologically based solutions difficult, if not impossible, and without a common language, we are left with a lot of statistics and hand-waving, but little real progress.

Our language reflects our thinking, and our thoughts determine the kinds of options we can imagine to the challenges we face in life. If talking about what comes out of human and other animal bums is linguistically problematic, then, Houston, we have a much bigger challenge to deal with than simply one of better engineering technology.

So let us step back for a few minutes. What is this stuff we are trying (not) to talk about? The simplest way to think about it is this: excrement is whatever your body doesn't use of the food it takes in, plus millions of bacteria that grow in your gut, plus quite a few of the cells from your gut lining. More specifically, excrement is defined in terms of an anal sphincter; it is defined by the animal it is leaving behind. Hence we speak of cow dung and baby poop, otter spraints and dog turds.

But how do we talk about it more generally? Are there any good words to use in semi-polite conversation? According to founder of the World Toilet Organization and global toilet guru Jack Sim, excrement — derived from the Latin word *excernere*, "to sift" — is the proper word. Certainly excrement is more acceptable in polite company than some of the alternatives. I am not so sure, however, that this is always the correct word to use for what comes out the anus, given all the other options available. In fact, I am not so certain that we should be agreeing on *one* term. Maybe we should agree on a family of understandable words.

What better way, for instance, to describe vapid, loudly proclaimed truth-indifferent verbiage than by referring to it as bullshit? Or, in more polite society, one might use the anglicized Dutch term for soft feces, poppycock (*pappekak*) — although the latter could, in the United States, be confused with Poppycock, the trademark for a type of candied popcorn much beloved by children.

And how else can one potty train a child than by speaking of poo-poo, which when used as a verb, is an onomatopoeia (a word that sounds like what it describes, and may have originated from a word that meant the sound of a horn blast), or doo-doo, or even number two? The numbering system, I hasten to add, is culturally relative. Writer Bill Bryson, when asked by a grade-school teacher whether he had to go number one or number two, exclaimed that he needed to have a big BM (bowel movement), which could be as large as a three or four. Little boys can joke about cow pies, cow pats, road apples, or turds, but what happens when they grow up? Perhaps there should be a globally standardized metric scale that runs from one to ten; this would enable global comparisons and further the science of excremental studies. In the mid-1980s, I recall that the attendants at outhouses in Borobudur, Indonesia, required patrons to declare before they entered whether they were going to do something big (take a dump) or small (urinate). This was based on an honor system, and the entrance fee was scaled accordingly.

The word "guano," used to describe feces from bats and birds, comes to us from South America via the Spanish. It is derived from the Quechua word *wanu* meaning fertilizer, and has a turbulent history related to both fertilizer and bombs, to which I shall return later. "Ordure," from Middle English, and before that Old French (*ord*, meaning filthy), derived from the Latin *horridus*, carries unhelpful (at least for agriculture) moral baggage, although, taking a leaf from William Blake's *The Marriage of Heaven and Hell*, one might re-frame the origins not as *ord*, but as *or*, which means gold. Where public order and religious ordination fit into this picture I shall leave for the anthropologists to pronounce.

More recently, the terms "biosolids" and "sewage sludge" have made their appearance in the technical and government literature to describe solid organic waste. They do neutralize the moral sting but are too technically scientific to be of much use in more general conversations. On the other hand, the term "nutrient," used by some agricultural bureaucracies, as in the phrases "nutrient cycling" and "nutrient management," is too general, and could just as well refer to the fat content in an avocado as the fat in a spraint. These phrases highlight the fact that excrement is comprised of useful nourishment for some species. However, the bacterial diversity in feces, the myriad roles it plays in ecological functioning and human survival, and the complex diversity of plant and animal life — which is the glorious visible manifestation of this nutrient cycling — are rendered invisible.

"Night soil" and "humanure" are attempts similarly to

neutralize the language of human shit, or perhaps even put a positive spin on it, but have yet to enter (or re-enter, in the case of night soil) general usage.

"Frass," which comes from the Old High German *frezzan*, to devour, is used to refer to both the excrement and the uneaten debris left behind by insects. Some have argued that the term "fecula," coming from the Latin *faecula* (crust of wine) and *faex* (dregs, sediment), should be used for true insect excrement, saving "frass" for the debris and refuse left behind by boring (albeit interesting) insects. Lest the reader think that entomologists have too much time on their hands, that they are arguing over how to name insect poop, I should remind them that science has conventionally progressed by making fine, precise distinctions (see discussion of shit, below), and puns such as are indulged in by this writer are generally frowned upon.

The vulgar term "crap" is sometimes attributed to the ingenious (or perhaps disingenuous) Thomas Crapper, who popularized the flush toilet in the late nineteenth and early twentieth century. The idea of flushing away excrement, however, can be traced back at least to the fifth labor of Hercules (or Heracles, as he is often referred to), a story to which I shall return later. The word "crap," while capitalized on by Mr. Crapper, can be traced back to Middle English (*crap* meaning residues from rendered fat), Old French (*crappe*, or residue), and medieval Latin.

An archeologist might speak of coproliths (hardened fecal balls in the bowels) or coprolites (which, despite the low-calorie name, are actually petrified shit-balls). Zoologists learn about animals by looking at scats and studying scatology; physicians examine stools (the word being derived from the place where the poop is delivered from the body) and fecal samples, a repertoire expanded by veterinarians to also include the examination of droppings.

Like many of the words in this linguistic landscape, "shit" encompasses all the schismatic contradictions of our cultural-scientific divides, and is thus problematic as a term to enable us to work toward resolutions. Still, its very protean nature suggests a cultural resonance not found in the other words.

Shit is used as an expression of dismay and disgust (a piece of shit) or frustration (oh shit), surprise or incredulity (no shit?!), or to describe trouble (in deep shit, up shit creek without a paddle), casual conversation (shoot the shit), cowardice (chickenshit), fear (shit one's pants), hysteria (apeshit), insincerity (horseshit, bullshit), care (to give a shit), anything that one doesn't like (looks like shit, tastes like shit), or substances, particularly illegal drugs, one likes (best shit I ever had). It is also one of the few words in the English language where the noun ("a pile of shit"), all tenses of the verb ("he shit, he shits, he will shit"), and the adjective (shitfaced), are all the same, a characteristic it has in common with the F-word. Andreas Schroeder, in his remarkable prison memoir *Shaking It Rough,* heard a fellow prisoner curse a broken machine by exclaiming, "The fuckin' fucker's fucked, fer fuck sakes!" Using the S-word in the same situation, however, would not have the same *je ne sais quoi.*

Like its sibling scat, shit has the same ancient proto-Indo-European root (*skei-*) as the word science, with a meaning having to do with separating one thing from another. This gives us the expletive "Scat!" used to shoo away wayward animals or children, or the staccato improvisations of jazz scat-singing, the scattering of seeds, the separation of dung from the anus, and the notion that someone is scatterbrained. These Indo-European roots are reflected in the Greek *skhizein* and Latin *scire*, which refer to cutting and splitting (hence telling one thing from another, as in "science," "conscience," and "conscious"). These words are all similar to excrement in their emphasis on "ex," something that separates from its origin.

This ability to separate reality into manageable pieces is at the root of the astounding success of industrial societies and modern science, and represents its greatest weakness. The strengths of disassembling the world and specialization are obvious: by understanding the bits (chemicals, bacteria, car axles, neural connections) we have accomplished amazing feats of chemistry, microbiology, industry, and neuroscience. The weakness has only become apparent after several centuries of success; indeed, the weakness may only be a weakness because of our success. We have saved so many babies, fed so many people, built so many cool cars that we are putting the whole globe at risk.

This weakness, a fundamental challenge for anyone talking about "sustainability," and reflected in the inability of cultural innovators (poets, novelists, musicians) and scientific investigators (biologists, physicists, coprologists) to speak comfortably with each other, to easily cross-reference each other's work, and to account for each other's evidence, is one of the central problems of the twenty-first century. We can assemble cars from parts created from metals mined in China and carbon-based liquids wrung from tar sands in Canada, and we can deconstruct and reconfigure genetic material to create tomatoes that live (albeit tastelessly) almost forever, but we cannot imagine our home in the universe. We can make heart-rending music, tell engrossing stories, and celebrate inspiring rituals, but, at anything other than an intellectual game level, are unable to relate this to our biological selves, our chemistry, and the physical structure of the universe. Ultimately, excrement is about all of these things.

Retired marine biologist Ralph Lewin has compiled a comprehensive summary of all things excremental, which he cleverly titled *Merde: Excursions in Scientific, Cultural, and Socio-historical Coprology*. This evades the problem in English but simply (as is the case with many solutions to environmental and cultural problems) displaces the problem elsewhere, a kind of linguistic NIMBY.

Failing a good English word that encompasses dung, conscience, consciousness, excrement, and wholeness, and to underscore the fact that words are not the ineffable reality they seek to describe, I am partial to a wide array of descriptors for what passes from the bodies of animals, rather than one definitive representation. After all, like the idea of a Great Being (male, female, spirit), the substance we are talking about has many faces and plays many roles in nature and culture. In keeping with this, I have tried in this book to use the terms most appropriate to particular fields of inquiry (manure, dung, ordure, frass). I have also tried to vary the words I use, so as not bore you (or myself).

In the end, however, I often return to using the word shit, impolite and problematic as it is. It is one of the few words that storms successfully through all the artificial barricades we have erected to block the streets and alleys between the deodorized proletariat and the sanitized ruling class, between popular and academic culture, between science and everyday life.

A few years ago, a fellow in the United States was taken to court and charged with obscenity because a righteous fellow citizen complained that the "SHT HPPNS" on his vanity license plate was offensive, to which the defendant replied that it stood for SHOUT HAPPINESS; what had the complainant thought it meant? So it seemed to me that if I use the most common word, and sometimes accidentally leave in the vowel, you will know what I mean, and perhaps not be offended, and if you are, well, SHT HPPNS.

CHAPTER 2

THE LIST OF INGREDIENTS AND AN INVENTORY

Before we go too much further, we should probably ask a basic question: what, really, is shit? I mean, what is in it? What are the ingredients? Only with that answered can we begin to explore some possible pathways out of our current dilemma.

From your lips to your butt, your intestines are a tube through which the environment of the outside world moves through your body. This is true of most warm-blooded animals. The muscles and organs inside animal bodies, including those of people, are sterile, and the interactions between the environment moving through the gut and those sterile organs are carefully regulated. That is why surgeons have to wash their hands and wear masks. If an appendix ruptures, or the surgeon nicks the intestine, you are in big trouble. The bacteria and food wastes in the intestine spill into the sterile field. This is why, from a food safety point of view, a rare steak is pretty safe; the inside of the muscle is sterile and you just need to burn the bacteria off the surface.

Eating is to the animal-environment relationship what sex is to relationships between animals: food represents the Earth inside you, becoming part of you. If we were so bold as to look inside our bodies or to travel there on an incredible journey, we would see the individual cells draw in oxygen and various elements and excrete what they don't use. The cells lining the intestine try to keep out as many toxins as they can and only absorb what is necessary for fueling the other cells that make up our bodies. Our own cells produce waste that could kill us if it were not carried away in the blood and then out of the body through urine, shit, bile, sweat, and breath. Some substances undergo special processing in the liver or kidneys before they are released. On top of this, every day tens of billions of our own cells commit suicide, an act referred to scientifically as apoptosis, or programmed death. Other cells, especially those that line the intestine, are scrubbed away by the passage of food and come out in our feces. Every day, parts of you are replaced. Have a look in the toilet or chamber pot. That's not just a stool sample: that's the old you. That's life.

Many of the materials in the food that are not allowed across the intestinal walls into the more intimate secret places of your body are used as food for the trillions of bacteria that are born, live, breed, and die in your gut. Mostly these are beneficial bacteria, which do their best to keep out the pathogenic bacteria, the disruptive hooligans of the bacterial world. Some of the wastes from these bacteria also provide further nourishment for your body in the form of certain vitamins. Eventually, this whole mass of undigested food materials, dead bacteria, toxins, and various fluids pumped across the intestinal wall pass through the body to be returned to the earth. This is what we call excrement.

You are part of a larger living web. Every life form, as the result of consumption and respiration, uses nutrients and energy from its surroundings and (re)cycles various kinds of modified or unused nutrients and energy back to the environment around it. Some of these byproducts of life are toxic to the organisms that dump them, some toxic to competitors or predators, and some are merely not used, for a whole variety of reasons. They may not be digestible by the animals that ingest them. Or they may be more than the animal can digest at a given meal.

Babies are born sterile inside and out and are only colonized by bacteria as they emerge into the wider world. Most of the hundred bacterial species, representing forty genera, that colonize a newborn baby's intestines in the first few months are either harmless or beneficial. As we get older, we acquire more bacteria from the parts of the environments we eat or otherwise interact with. There are somewhere in the vicinity of 10^{11} bacteria (that is ten with eleven zeroes after it) per cubic millimeter of adult human feces, a number that exceeds my capacity to imagine, but is typical of other animals' feces. These bacteria represent some 500 to 1,000 different species, most of which we don't know much about. These include a wide range of micro-organisms, including bifidobacteria and lactic acid bacteria (which predominate in breastfed babies and yogurt), coliforms (which scientists look for when measuring water contamination), and even archaea (which prefer an atmosphere devoid of oxygen, much like that of the early Earth).

Most of the micro-organisms that live in animal intestines, including *Clostridium* species (which can, under the "right" circumstances, cause serious diseases like tetanus and botulism), are beneficial for the health of both the host species and the ecosystems in which they live. In a 2011 review of intestinal microbiology, scientists Reading and Kasper wrote, "While mammalian hosts provide a nutrient-rich niche for these bacteria [that live in their intestines], the bacteria provide the host with much more including: aid in digestion, protection against pathogenic (disease-causing) enteric pathogens, and development of the immune system." The beneficial relationship of these bacteria for immune system development seems particularly important. Without bacteria in our intestines, we would waste away and die. Indeed, if some evolutionary biologists are to be believed, the cells that comprise our bodies, and the organelles and mitochondria within them, are communities of co-evolved bacteria. Without bacteria, we literally wouldn't exist. We *are* bacteria. This book may have been written by a community of bacteria trying to understand itself. What a wonder!

From another, non-microbial, perspective, excrement may be described in terms of its chemical composition. This is the viewpoint generally taken by those interested in the sustainable management of manure; in particular, the chemicals in excrement may be thought of as nutrients for other species, and as fertilizer supplements to soil. The nutrient content of manure, like the labels on canned foods, tells us a great deal not just about what is being eaten, but also what we might be able to do with the leftovers.

These contents have been intensively studied in farm animals, in part because there is an economic incentive for livestock producers to ensure efficient use, and reuse, of resources. For instance, whatever its challenges with regard to palatability, the addition of chicken manure, with 26 pounds per ton[1] of nitrogen to cattle feed, is a way of adding protein to the diet. The nitrogen in chicken shit can be used by the bacteria in the ruminant stomachs to produce protein. One of the reasons that bird manure in general is so high in nitrogen is that, in birds, manure and urine come out through a common opening, the cloaca, in a mixed slurry.

Mammalian urine is also a very "hot" source of nitrogen: for verification of this you need only look at the brown areas of grass where the dog has urinated, killing the plants with too much nitrogen. For this reason, some investigators have argued that, for people, solid excrement and urine should be separated at the source, with the urine (which is mostly sterile) going directly into use as fertilizer, and the solid waste (which is where the bacteria live) going to composting or bio-digesters, which we shall consider in more detail later. There are toilet designs that can accommodate this, although livestock manure management schemes have tended to err on the side of combining the two, in the name of efficiency. We shall discuss the problems that have accompanied the obsessive deification of efficiency in our society later, when we consider responses to our current challenges.

Another important mineral often deficient in soils and present in manure — although not in such high levels as nitrogen — is phosphorus. Cattle manure has only a couple of pounds per ton of phosphorus, pig manure about twice

1 A metric ton is 1,000 kilograms, or about 2,200 pounds. An imperial ton (sometimes called a short ton) is 2,000 pounds. Given the wide levels of uncertainty around all of the numbers used in this book, and the fact that we are interested here in orders of magnitude rather than exact measurement, I have treated the two as essentially equivalent when drawing on the research for this book.

that, and poultry shit is up to ten times that. They all have about seven to ten pounds per ton of potassium.

According to the authors of the Food and Agriculture Organization's report *Livestock's Long Shadow: Environmental Issues and Options*, the cumulative global manure output from livestock included 135 million tons of nitrogen (with cattle, at 58% of that number, being the biggest donors) and 58 million tons of phosphorus. Calculations on the economic value of manure are not standardized, as they depend on the local (and changing) ecological and economic contexts, but a 2001 report estimated that, in the United Kingdom, manure from dairy and beef cattle represented about 280,000 tons of nitrogen, 50,000 tons of phosphorus, and 250,000 tons of potassium annually. Poultry manure represented 100,000 tons of nitrogen, 40,000 tons of phosphorus, and 50,000 tons of potassium. The value of the poultry manure alone in the U.K. was estimated in 2001 at about £50 million (about US$80 million) per annum.

Human manure appears to be more variable in content than that of other animals, perhaps because our diet is so varied. There are many ways to calculate the content, and the literature on this subject is (to me anyway) confusing. Some authors report percentages, some weights, some dry weights, some wet weights, some in grams, and some in pounds. Next time you are staring out at your parched lawn, ponder these ballpark figures: our shit is 75% water. Beyond that, the 150 grams of daily output include on average 10–12 grams of nitrogen, 2 grams of phosphorus and 3 grams of potassium.[2] Although most carbon is dunged out (the carbon includes cells from the walls of the intestines in

2 These numbers are rough estimates based on "average people" without diarrhea from a variety of studies in different parts of the world.

large numbers — sometimes more than half of the bulk — of emigrating bacteria), people mostly piss their nitrogen and potassium away. Phosphorus is equally distributed between urine and feces. Our excrement also includes 8% fiber and 5% fat. Again, this may be in the form of partly digested food, bacteria, cells, and so on.

If we stay with the nutrient-chemical perspective, human shit has been calculated to retain 8% of the caloric value (a universal measure of energy) of the food we eat, varying according to our diet. We shit out 25% of the protein in rice, 26% of the protein in potatoes, 40% of the protein in cornmeal. A person could probably get by eating human excrement, but one would have to eat a lot of it in order to get the required protein and energy intake.

A research study on human feces in Thailand found that its chemical composition (elements such as nitrogen, phosphorus, potassium, calcium, magnesium, copper, and the like) did not vary significantly by age, sex, occupation, or religion. From these findings, one can infer that, in people, the characteristics of the shit producers are irrelevant to the quality of the shit they produce, although some might dispute the matter.

When we scan across different animal species, we see that not all manures are equal; when used under intensive conditions they need to be titrated as carefully as any commercial fertilizer. Every shit is different, every soil is different, and each crop has different nutrient requirements; most farmers in North America have their soils tested in laboratories before adding fertilizers. The challenge for using manures is that their exact make-up varies, so that in farming systems where nitrogen and phosphorus content of the fertilizer is expected to be precise, manures would seem to be less useful.

Another way to look at excrement, which has gained

greater prominence as the costs of fossil fuels have risen, is in terms of its energy content. I never actually thought much about the nutrient or energy content of manure until the summer after my first year in veterinary college. I worked for an applied physiologist who was treating straw with nitrogen (ammonia) and feeding it to sheep. The rationale for this study was that the bacteria in the stomachs of ruminants, like sheep and cows, can take cellulose from straw, break it down, take nitrogen from ammonia (or chicken shit, for that matter) and reassemble it; taken up by bacteria, these apparently useless inputs become protein for the sheep. The thinking behind this experiment was that feeding this treated straw to sheep and cows might offer all kinds of "value-added" possibilities for prairie agriculture, such as feeding "useless" straw to produce "useful" sheep.

My job was to take the treated straw and test its total energy content (the number of calories it contained) in a so-called bomb calorimeter, and then take the sheep droppings — essentially what was left over after the straw had passed through the sheep's complex digestive system — and test them as well. The difference in the energy content between the straw and the droppings (or, more accurately, the difference in their capacity to produce heat) would tell us how much net energy the sheep were actually gaining on this diet. While the premise was scientifically sound, the sheep behaved as if someone were trying to feed them straw that had been pissed on, which is sort of what we were doing. They seemed to prefer eating the wool off each other's backs to eating the treated straw. The moral of that story was that net energy isn't everything. Sheep, just like people, do not eat merely to gain or lose weight, or to bring in exactly the right amount of protein and minerals, or to avoid disease, but also because certain foods *taste* better than others, a fact seemingly lost on some laboratory-trained nutritionists and public health workers.

Describing the content of feces, whether in terms of bacteria, chemicals, or energy, is but a first tiny step in understanding what excrement is, and how we might better manage our relationships with it. For this managing what is important is not the thing-in-itself, but the thing-as-a-function, the relationships it represents, the cultural and ecological webs that give it meaning. Without that broader understanding, we literally don't know shit.

Consider, for instance, the story of guano, which underscores the importance of context as well as any. Birds and bats produce feces called guano, which is rich in ammonia, uric, phosphoric, carbonic, and oxalic acids, as well as nitrates. Bird and bat shit contains a lot of interesting chemicals. So what?

Robert B. Marks, in his book *The Origins of the Modern World*, attributes the explosion of the global human population in the eighteenth and nineteenth centuries at least in part to the discovery of major reserves of bat guano. Bat and bird guano were discovered by Europeans in the early nineteenth century to be an excellent source of saltpeter (potassium and sodium nitrate), useful for both fertilizer and explosives. Although it had been used in South American cultures for a millennium or more to enrich soils, the European "discovery" of guano's strategic value, resulted, after the 1840s, in rapidly increasing, unsustainable exploitation of its main sources. Bird shit from dry climates where the nitrates had not been leached out by rains — such as parts of Peru and Chile — was considered the best. Britain and its allies exploited Chinese labor to extract valuable guano from those countries to replenish their own soils. Spain fought Chile and Peru over access to guano in the Guano Wars of 1865–66. In 1856 the U.S. Congress passed the Guano Island Act allowing U.S. citizens to claim unnamed or uninhabited islands covered in guano for the United States. More than fifty islands were claimed for the U.S. under this act, including Midway Island in the Pacific — and an island in Ian Fleming's James Bond novel *Dr. No.*[3]

Although the nitrates found in guano have also been used to produce explosives, the amount of manure required to get enough ammonium nitrate to create a bomb is considerable and not usually considered practical. Attempts to do so have sometimes taken on the trappings of a *Saturday Night Live* skit. In 2008, two women in Germany reportedly slipped into an animal slurry tank while trying to fill their stockings to make "manure bombs." They fled naked, on foot, after one of them slipped into the tank. Without certain kinds of agriculture and weapons of war, guano is just shit. Context is everything.

If excrement were merely a valuable resource for agriculture and war, a kind of brown gold, we would probably just let the stock market brokers in New York and Toronto and London and Hong Kong haggle over how much money they could get trading it. We could invest, retire (or run for president), and smugly watch the share prices go up. Unlike gold or oil, however, the wicked shit dilemma is not related to scarcity, but to abundance.

The world is full of shit, and the amount appears to have been increasing rapidly over the past few hundred years. First of all, we should ask: is this true? And secondly, if true: how can this be? Where is all this apparent increase in excrement coming from? Surely the amount of

3 Guano was also taken up on Gemini and Mercury space missions and was used as the propellant to deploy the radio transmission antennas after splashdown.

matter on the planet has not increased dramatically in the past century. And although many tons of cosmic dust and rocks fall to Earth each year, they are not all transformed immediately into excrement.

Jack Sim, founder of the World Toilet Association, informed me in an email that the average person defecates once a day, eliminating 120 to 150 grams (a bit more than a quarter pound) of feces and 1.2 liters (about a quart) of urine. As we can attest from personal experience, the exact amount we produce depends on the volume and content of our food and water intake, and on our health status. Vegetarians produce more poop than meat eaters, because they ingest more indigestible fiber, and whether a person is constipated or has diarrhea also influences the output.

A scientific study published in 2001 in the *Science of the Total Environment*, using subjects in Thailand ranging in age from eleven to seventy-one years, found that each person generated 120 to 400 grams of wet feces and 0.6 to 1.2 liters of urine per day. These averages are in the same ballpark as those from studies done earlier in many other parts of the world, demonstrating that the Thai people are no more full of shit than anyone else in the world (which my Thai friends could have told us without a scientific study).

So, for argument's sake, let's say each person (the average one, somewhere between starving children in the Sudan and obese adults in the U.S.) puts out 150 grams of excrement per day. That's about 55 kilograms in a year. This is likely an underestimate, but at least it gives us a reasonable number to work with.

In the mid-1990s, while working in Kathmandu, Nepal, I came across a public facility labeled with a sign outside identifying it as "Boudha Toilet." This was a humbling reminder to me that, whatever our most lofty spiritual aspirations, we are all, also, animals. From this, we can conclude that Jesus, who lived about thirty-three years, dumped nearly two metric tons of poop on the planet during his lifetime. Mohammed, who probably lived about sixty years, processed enough food to put out nearly twice Jesus's total. Karl Marx upped Mohammed by five more years' worth of stool — about another 275 kilograms. And Buddha, at eighty years, trumped them all at more than 4,000 kilograms, which may partly explain the size of that large banyan tree in India under which the Buddha sat for so many years. An old Persian saying claims that the purpose of the human body is to turn the fine wine of Shiraz into urine. I have never heard a similar aphorism about the fine dates of Persia and feces, but such a saying very well could have existed.

More relevant to our discussion on the global volume of excrement is the rise in population. In 10,000 BCE there were about a million people on the planet. That's 55 million kilograms of human excrement scattered around the globe in small piles, slowly feeding the grass and fruit trees. In 1800, there were about a billion people on the planet, so about 55 billion kilograms. By 1900, we had a world human population around 1.6 billion, which would have been 88 billion kilograms of human shit.

By 2013, with more than 7 billion people on Earth, the total human output was close to 400 million metric tons (400 billion kilograms) of shit per year. That is about 80 million large bull elephants' weight of crap! Okay, I can't imagine that either.

Now let us consider all the other animals with whom we share this planet. Many are going extinct at dizzying rates, but this does not mean that the overall number of animals is decreasing. For instance, beetles, bats, and frogs are disappearing, but the numbers of commercially raised pigs and chickens are dramatically increasing.

No one seems to have dared guesstimate livestock numbers until the 1960s. Nevertheless, we know that most of the major cultures in the Middle Eastern cradle of civilization were cattle cultures, so there were no doubt, as some are wont to say, "quite a number." I grew up with an old hymn, based on ancient Jewish stories, that asserted God's ownership over the cattle on a thousand hills, but it didn't say how many cattle were on each hill. In any case, by 1961 there were said to be more than 900 million cattle in the world, as well as 400 million pigs, about a billion sheep and goats, and almost 4 billion chickens. By 2010 (the most recent figures available at the time of writing), there were about 1.4 billion cattle, 19 billion chickens, a billion pigs, and 1.8 billion sheep and goats.[4]

So we know there is a lot of animal dung in the world, and the amount is increasing. How much is a lot, though? Let's be really rough about this. I mean, who is actually counting individual samples of chicken shit globally anyway? People who give you exact numbers are engaging in false precision, which is impressive for bamboozling bureaucrats and politicians but not of much use otherwise. As long as we are within an order of magnitude — that is, are we talking thousands, millions, or billions of tons? — I think we'll get the general picture.

Livestock rearing is still on the increase, rapidly, most rapidly in developing countries, and mostly involving a

4 What this count doesn't tell you is the turnover — how many billions of chickens are born and killed each year, for instance. Nevertheless, these numbers from the United Nations Food and Agriculture Organization will give us a rough idea for these calculations. Between 1961 and 2010, sheep numbers remained steady, but goat numbers went up.

major shift from conventional mixed farming to industrial agriculture. If we look at the countries with the largest numbers of livestock in 2010, we can begin to get a sense of recent growth and where it has occurred. Between 1961 and 2010, the estimated number of chickens in China went from 540 million to 4.8 billion, the number of pigs went from 380 million to 476 million, and sheep went from 110 million to 134 million. If we lump cattle and buffalo together, India (with a pretty even mix of both) had the biggest numbers, going from 226 million in 1961 to 310 million in 2010; Brazil (with mostly cattle) went from 56 million to 210 million.

There are a few ways to calculate how much manure is produced by livestock, usually based on "average" animals, which only exist in the phantasmagoric world of statistical charts. A high-producing dairy cow in Wisconsin is quite a different animal from a garbage-browsing bovine in Bengal. Still, for our purposes here, averages are good enough, because the point I want to make is a general one, and not intended for use by farmers for manure-management programs. Using low-end estimates of fecal output from current livestock in Canada,[5] I arrived at the following: in 2010, the amount of manure produced by all the cattle, sheep, goats, pigs, and chickens in the

5 Statistics Canada estimates: bulls (42 kg/day), beef cows (37 kg/day), steers (26 kg/day), heifers (24 kg/day), and calves (12 kg/day). Milk cows produce the most manure at 62 kg per day. I picked 24 kg to keep a low-end estimate. Pigs including weaners, sows, boars, and market hogs produce between 1 and 4 kg per day. I picked 2 kg to stay at the low end. Poultry produce less than 1 kg of manure per bird per day. I used 0.07 kg/bird, which I estimated from various agricultural extension sites. For all animals, I aimed for the low end. No point in exaggerating an already big number. See Statistics Canada, "A Geographic Profile of Livestock Manure Production in Canada, 2006" at http://www.statcan.gc.ca/pub/16-002-x/2008004/article/10751-eng.htm#a4.

world added up to 14,136,450,000 metric tons (multiply by 1,000 to get kilograms). That's 35,341,125,000 million cubic meters.[6] If we take an average soccer field to be about 60 meters wide and 100 meters long, we could cover almost 3 million fields to a depth of 2 meters — enough to bury everyone on the field. (I left out the human manure, because we want somebody to be able to watch this spectacle.) If you want a more homey way to think about it, that's 141,365,000,000,000 one-cup servings — enough for 2,000 one-cup servings a year (more than five cups a day) for every person on the planet.

And then we can throw in the dogs and cats and rats and coyotes and the deer and elephants and robins and pythons and the bird shit on your car's windshield. Animals that are well adapted to human settlements, such as cockroaches and raccoons, are also increasing, although these are not as carefully counted, and cockroach frass is not easily quantified. Overall, I am guessing we are facing a grand total weight of about 400 million tons for people, and more than 14 billion tons from the other animals — every year. And the amounts are going up, even as some economists are telling us we need more people to keep the economy going. Let them eat shit, I say. Well, maybe they will, but we'll come back to that later.

In any case, I do not think we need an international scientific study to tell us that there is a whole lot more excrement in the world than ever before in history.

So how did we get to this point? Perhaps if we go right back to the beginning, to the origins of excrement, we can decipher some clues to find a way ahead.

6 To calculate volume (cubic meters, cups) from weight, I used the online ConvertMe tool, which assumes a manure density of 0.4 kg/liter. A metric cup is 250 mL.

CHAPTER 3

ON THE ORIGIN OF FECES

Before life, there was no shit.

Some time more than 3 billion years ago, life came into being. Life as we know it emerged from pre-living, biochemically active ancestors, which, according to some to researchers, lived in porous rock pockets near warm, alkaline vents in the sea floor. In this version of the story, these porous rocks provided small cave condominiums (like hollow grapes); the stone walls were the boundaries within which closed loops and cycles of biochemical interaction could develop.

Many of the molecules that make up what we call life — organic molecules such as carbohydrates and proteins — formed as the result of chemicals reacting with each other in a warm, energy-rich soup. You can have lots of chemistry without life; you can't have life without chemistry.

Stuart Kauffman, the world-renowned complexity theorist, argued in his book *At Home in the Universe* that the "poised biosphere creeps upward in a controlled generation of molecular diversity, not a vast wild explosion. The wild explosion would occur if . . . all cell membranes disrupted. Cell membranes block many molecular interactions and hence block a supracritical explosion."

Without membranes, there is no shape to this orgy of interactions, and without shape, there is no life. For all of us, having skin is essential to keeping in the fluids, the chemical reactions, the cells, and the hormones that make us who we are. Without this skin, we leak out into the surrounding environment, and the chemicals in our bodies just wander away to react with other chemicals. We become one with the Earth — which is another way of saying we become dead. A dead skunk and a living skunk are comprised of the same chemicals. The difference is that the dead skunk's chemicals are uncontained. This necessity of having a membrane is true all the way down into the organs and cells that make up who we are.

At the beginning of all this, billions of years ago, the cloaking of these biochemical reactions resulted in the first single-celled organisms, whose existence we infer rather than have direct evidence for, precursors to archaea and bacteria. Neither bacteria nor archaea have bounded nuclei or organelles (membrane-bound mini-organs, which may represent what is left of even smaller microbial invaders) within them. Archaea seem to prefer more extreme environments, and have been discovered in boiling water as well as in water that is super-salty, near freezing, or close to sulfurous volcanic vents. Once thought to be precursors to bacteria, archaea are now understood to have arisen from a common ancestor, perhaps proto-living things in the porous rocks.

All living organisms, from single-celled bacteria to multi-cellular elephants and people, need somehow to capture energy, which then enables them to fuel the reactions that occur inside them and give them life. So, for instance (and I am simplifying several university courses, encyclopedic biology books, and many controversies over how to interpret new evidence uncovered every year), cyanobacteria, also called blue-green algae, take in energy from the sun, carbon from carbon dioxide in the atmosphere, and electrons from water to make sugars; oxygen is released as a toxic waste (that is, toxic to the cyanobacteria), a kind of gaseous turd if you will, and needs to be transported outside these organisms, or they will die.

In the early days of life on Earth, bacteria evolved that can use that expelled oxygen to generate their own energy, and can take nitrogen from the air and phosphorus from the soil to incorporate it into molecules — such as those comprising amino acids and DNA — that are useful for life. Without a cell wall, and with a high concentration of proteins and sugars inside, the cells would explode in more chemically diluted surroundings. If the liquids outside the cell have, say, higher salt concentrations than occur inside the cell, then, without the membranes and active pumping of various ions across those membranes, the organisms shrivel and die. Being bounded — with controlled leakiness — is essential for life. We think of a system that can exchange energy, information, and materials with its surroundings as an open system. The alternative is a system whose boundaries won't let anything in, or out. That would be a closed system. All organisms, including us, are open systems.

Some of the energy that comes into a cell is used to run activities like the sodium-potassium pump, which helps keep potassium high and sodium low inside the cell, and also enables transport of other molecules across the cell wall. Some of the energy may be used directly to create proton gradients, which then can be used to run, say, a spinning flagellum, or tail. This can enable a single-celled organism to get around, like a tiny self-propelled motorboat. Much of the incoming energy, however, is stored in chemical bonds that form molecules. Think of them as secured government savings bonds — or maybe, these days, money under the mattress. The bonds, when broken, release that energy. Adenosine-5'-triphosphate (ATP), which contains nitrogen, phosphorus, and oxygen, is one of the most common of the energy-storing molecules. This energy can be moved around and used later.

Energy is constantly bombarding Earth from the sun, supplemented by energy from volcanic and other geothermal sources. The high energy outside the cells from heat and sunlight creates a gradient that stimulates the creation of complex structures as a way of dissipating, or using, the excess energy. Otherwise the organisms would simply boil to death. Thus organisms developed not just storage chemicals, but also networks of cell-bound organelles and strands of DNA enclosed by membranes (plasmids in bacteria, the nucleus in eukaryotes). For this reason, some complexity theorists have argued that life, as a complex system, only emerges and exists far from equilibrium; that is, if everything were in equilibrium, there would be no energy gradients across organisms' walls, and the amount of disorder (entropy) would be the same inside as outside. There would be no life. To badly paraphrase Leonard Cohen: "Life is bounded and porous. That's how the energy gets in."

Enclosed by permeable membranes, with useful chemicals coming in and unused or toxic products leaking or being pumped out, the reactions in these ancient cells began relatively simply. As they created structures to deal with all that incoming energy, they evolved into something considerably more complex.

In the long story of life on Earth, it appears that the organisms that can't live with oxygen (the anaerobes), and hence produce it as waste, preceded those that need oxygen (the aerobes). Until about 2 or 3 billion years ago, the atmosphere was dominated by carbon dioxide (90%) and other gases that we now consider toxic, including methane and sulfur. Those early organisms took for granted that the atmosphere would stay that way as long as the sun shall shine. This is another way of saying that only those who could thrive and reproduce in such an atmosphere survived. Some ecological theorists have phrased this in a way that suggests the organisms anticipated that the future would be like the past. Others, like philosopher Karl Popper, described this as a kind of "propensity" in organisms and larger systems to keep going in certain directions. However one describes it, this "taking-for-granted anticipatory propensity" only works if tomorrow is more or less like yesterday. The Earth is littered with the fossils of those who made such assumptions.

A couple of billion years ago, as the "toxic" oxygen waste in the atmosphere built up, and carbon dioxide dropped (today it's about 0.02%), these early anaerobes "disappeared" from view. Trillions died.[1] But they didn't *all* die. Some survived, and are still thriving, in the many places on Earth where oxygen is absent or rare — deep in the soil, the water, and in every animal intestine, including your own. Although anaerobes sometimes cause serious

1 The periodic mass kill-offs of people and other animals from "acts of God" over the millennia lead me to ponder the veracity of bumper stickers asserting that "God is pro-life." If so, S/he has an odd way of showing it.

problems (botulism and tetanus are both caused by anaerobes), they are still essential for life as we know it.

Over time, and differently from place to place, organisms emerged that could use the waste of other organisms, usually after it was transported outside the cell wall. Some microorganisms moved inside others, closer to where the "waste" was being produced. These adapted to a kind of cousin-who-refuses-to-leave apartment life, forming symbiotic relationships, so that both the "parasite" and the "host" benefited. We tend to think of lichen, which are a combination of algae and fungi, when we think of symbiosis, but we could also think of the cells that make up our own organs and bodies as living symbiotically. According to one version of early events, the mitochondria inside each of our cells, which make us possible as multi-cellular animals, are probably descendants of bacteria that moved into archaea.

In other words, individual organisms — plants and animals — began to interact in complex ways, becoming dependent on each other, and in the process creating what we think of as ecosystems. The notion of ecosystems is controversial among some ecologists, since, unlike organisms, they have no visible membranes or genes. This is complicated by the fact that many scientists, trained to think in an industrial linear fashion, have a difficult time conceptualizing ecosystemic development over time as anything other than an artifact of natural selection on organisms.

Here, and throughout this book, I shall speak of ecosystems as if they exist as "real things." Please note that I am ambivalent — one might even say agnostic — about this. Nevertheless, the language of ecosystems provides a shorthand, useful way to talk about the multiple, interacting pathways of nutrients, energy, and information (in the form of seeds and other packets of DNA and RNA), and their collective (emergent) behavior. Whether or not ecosystems exist as a physical reality, the language of ecosystems, and the mental models this conjures up, provides some useful tools for understanding the complex world we inhabit. Imagine, for a moment, that you are looking through a microscope at cells of various sorts. When you step away from the microscope, something else emerges. You see that the cells are actually parts of larger things; instead of seeing cells, you see plants and animals. Now, imagine that you are a giant who sees plants and animals (including ourselves) as squiggles on a microscope slide; when you step back, instead of seeing plants and animals, you see ecosystems. Practice this a few times. The imaginative skill required to move from cells to organisms to ecosystems (and beyond) is important for understanding the multiple, nested connections that will help us grapple with the wickedness of shit.

These ecosystems, with somewhat leakier boundaries than the organisms of which they are comprised (think of coral reefs, or boreal forests), developed their own sets of internal and external rules and structures regarding energy, material, and information exchange. The species that make up a given ecosystem, and the nature of their relationships, depend on history (both in terms of human activities and which seeds and animals happened to be there at particularly opportune moments), the nature of the soil (including its resident bacteria), water availability, and climate. Hence we have boreal forests and coral reefs, semi-arid grasslands and cold marine systems, each with different species and different (and usually shifting) boundaries. Like the organisms of which they are comprised, these ecosystems are built around certain assumptions that the future will be sort of like the past; they do not anticipate incoming meteors or human population explosions or global warming or massive piles of excrement. They do not expect that their neighboring species will die off, or leave. They anticipate spring-summer-fall-winter, or dry season–small rains–dry season–big rains, or ocean currents from a particular direction at a particular temperature with fluctuations every five to ten years. These expectations are built into the relationships between their DNA and the contexts they live in.

Let me summarize this somewhat differently. Again, practice your ability to imagine the multiple layers of life that emerge when you view the world at different spatial scales. Individual plants and animals are like well-integrated communities of interacting single-celled organisms that take in energy, use it to create structure, and then excrete what they (we) don't need. As happens at the single-cell level, the energy excreted from one multicellular organism may still have considerable residual value for others. For example, plants pass on energy by being eaten, by dropping off parts like leaves, or by dying. This provides food for fungi, bacteria, and insects, a process that from the plant's point of view is decomposition, but that from the point of view of these tinier life forms, and, say, the birds that eat the insect larvae, is actually a process of composition and re-composition. The notes of the plant provide the raw material for the symphony in the compost heap. We (animals) also die and enter the same decomposition and re-composition cycles, but before we do, we eat a lot of plants and other animals and pass along a lot of nutrients and energy as shit that is still useful to other organisms, large and small. All of this activity is based on certain anticipations that the future will be at least sort of like the past, a kind of ecological stability. What emerges from all these interactions are what we call ecosystems.

Again, thinking small, we can imagine how multicellular organisms emerged from the orgies of single-cell organisms (who are considerably less hesitant to share bodily fluids and genetic information than people). As we stand back a little, we can see that, among the offspring of single-celled interactions, plants and animals represent two overlapping strategies to exchange energy, information, and nutrients with their neighbors and the general environment.

Plants, which are usually rooted in one place, select the best sources of energy and nutrients from the sun, soil, and water around them. Animals, which are usually mobile, move themselves to the best possible sources. Sometimes the boundaries between what we think of as plants and what we call animals are blurred. There are organisms such as slime molds that sometimes behave as animals, moving from place to place, and at other times grow as a plant would, in one place. Coral animals tend to stay close to home, and aquatic plants can float a long way from where they are born. Dinoflagellates live in marine environments and can create toxic blooms; some of them can use solar energy (through photosynthesis), while others are more like single-celled animals. They have been classified by both zoologists and botanists, as if by naming them we could make nature fit into categories. Mingling among these plants, animals, and plant-animals are viral and microbial populations, which travel both inside and outside plants and animals, and are neither.

The point to remember here is that any naming or

classification system, even one based on observable features, is ultimately a human construction for human use, and is dependent on, among other things, our scale of observation and our perspective. Waste and shit are categories relevant to animals and plants, but not to ecosystems, where nutrient cycling classifications are more important. This will become important when we begin to wrestle with what we classify as waste, sometimes call shit, and rarely think of as essential for the life of the biosphere.

Like the idea of shit as something separate from ourselves, the distinction between plants and animals is most useful at the ordinary, practical level of daily life where most of us focus our attentions, that is, neither looking through a microscope nor looking down from space. In our daily lives, we observe that plants do not technically produce waste in the same way that animals do. Furthermore, while anaerobic bacteria may be offended at the scent of oxygen farts passing from a bush, most of us relish the fresh scent of clean air in the woods at night.

Plants have evolved to take in energy from the sun and water and nutrients from their surrounding soil to create biomass. Because the source of these inputs either has to be where plants live or has to come to them, the cells in plant roots try to select molecules that are useful to them and leave the rest in the soil or water. Although plants do not have digestive or intestinal systems, their roots are in a sense like animal intestines turned inside out, with the root surfaces serving functions similar to those of the cells lining the intestines. The waste that plants excrete (through the leaves) is mostly oxygen (at night) and carbon dioxide (by day).

Since plants can only draw on the area immediately around them, it is often advantageous if their children move to a new neighborhood to avoid overtaxing locally available resources. Depending on where they live, and who the neighbors are, many strategies to deal with this local competition have evolved. Usually these strategies involve animals — birds, fish, wandering herbivores — and often entail seeds traveling in shit. This means that how we manage animals and their excrement has profound implications for plants as well. Again, understanding these relationships requires us to see the ecosystems that emerge from them, to keep making the dizzying transitions from cells to organisms to the webs of life in which they are embedded.

A wide variety of animals eat fruit and shit out the seeds, from bears and cassowaries to fruit-eating bats and fish, making forests of mixed vegetation possible, and transporting species over wide areas. Nomadic people have no doubt also filled this role in the past. Animal shit is essential for the spread and survival of many different plant species and hence for the types of landscapes people have inhabited.

Some plants have "conspired" with other species to limit the geographic range of the dispersal (keeping the kids close to home). For instance, birds that eat Australian mangrove mistletoe fruits pass seeds in less than ten minutes because of the laxative qualities in the seeds.

But not all seeds and young want to stay close to home. In some cases, seeds that stay too close to parent trees are competing for nutrients with their parents, and may succumb to diseases and pests that have concentrated around the parent plant (competitive clustering). Young animals can leave the nest and wander to find new territories and food sources. Plants are thus often dependent on animals and their shitting habits.

The science of tracing which seeds or spores came from which plants has waited for newer genetic techniques, and the more we learn, the more complex it looks. Birds and bats disperse seeds, as do a variety of plant-eaters like goats and deer, and small omnivorous mammals like foxes. Conserving the biodiversity of plant populations means we have to understand what foxes, deer, goats, birds, bees, and bats are eating, as well as how far they travel and where they shit. Nor is this intuitive. In at least some studies, the small mammals seem to travel farther than, say, bats in the same area, but drop the seeds on rockier, more open ground.

Some plants excrete toxins into the soil to keep away other plants. Others, having no excretory mechanism, must hold these toxins in their tissues in closed-off structures (vesicles) or (as they do with heavy metals such as cadmium) chemically bind them and try to keep them out of general circulation. Sometimes plants drop off infected leaves (a process called abscission), which is as close to shitting as a plant is capable. The fact that some plants hold on to toxic elements like cadmium or cesium-137 (the radioactive element that mimics potassium in biological systems) might seem of only tangential interest in a discussion of excrement. In the twenty-first century, however, when people harvest the plants, transport them somewhere else, and feed them to cows or pigs or chickens, who shit out what they don't use, the cadmium or cesium content of the plants is at least interesting, and perhaps worrying.

When plants die, they decompose, and bacteria can recycle the energy and nutrients (such as nitrogen that they have taken from the air, or phosphorus from the soil) plus any toxins (such as cadmium) back into other living organisms. Plants, and their waste products, are food for animals.

Ruminants (sheep, goats, deer, camels, giraffes, and the like) have a series of stomachs — the abomasum, omasum, reticulum, and rumen — that enable them to process plant materials much more completely than, say, a horse or a pig or a human. The rumen is a kind of swamp, teeming with about 200 species of bacteria and protozoa. Some of the anaerobic bacterial species can digest cellulose, which is why ruminants, unlike people, can utilize grass and straw as sources of energy; the protozoa eat the bacteria, die, and become food for the bacteria; all of them provide protein for the ruminant.

The importance of cellulose-digesting bacteria (which also live in termites and enable them to eat wood) goes well beyond their own cellulose-digesting abilities. These organisms take cellulose from plants, digest it, and, by combining it with nitrogen from almost any source (chicken shit, ammonia), create proteins. When these microscopic animals die, they provide protein to the cows and sheep and deer. This is why ruminants do not have the same dietary protein requirements that birds, people, or other non-ruminants do. This is also why I always find puzzling the charts comparing feed efficiency in, say, chickens and cows, in which the chickens always come off looking better. The ecological point is that cows can eat plant material that is simply not available for use to human bodies. Chickens, which do not digest cellulose, are competing with us for some of the same sources of proteins and carbohydrates: oilseeds and legumes like soybeans and peas, cereal grains like barley and wheat, and animal proteins like fishmeal.

Although waste usually emerges from the back end of an animal, this is not always the case. Jellyfish and other simple invertebrates regurgitate undigested materials, which is like shitting out your mouth. A variation of this also happens in some vertebrates. If you are a predatory bird like a vulture, owl, or hawk, or a shellfish-eating shorebird, or a sperm whale, the choice is whether to try to painfully push the undigested bones, beaks, fur, and shells through your unwilling intestinal tract or to regurgitate them. I confess, given the choice, I would also go with regurgitation, and then a gin-and-tonic chaser to cleanse the palate.

The cells of teleost fishes, coral animals, sea anemones, sea pens, urchins, starfish, and other ancient animals (ancient as in a couple of hundred million years, hence older than the so-called ancient mariner) release wastes — often high in toxic ammonia — into their internally circulating fluids. From there the wastes diffuse into the surrounding water. In this case, the historical solution to pollution has indeed been dilution.

Now stand back, again, from the details of ingestion, waste disposal, and movement I have described. The world we live in is not just a collection of trees, cows, and birds producing waste and shit and interacting with each other. What emerges from all these interactions, what we see when we stand back and imagine ourselves as giants, are what some of us call ecosystems. Being able to imagine these larger webs of which we are a part is essential for shifting from a mindset of "sustainable manure management" to sustainable living in a biosphere in which nothing is wasted. If we can envision these encompassing systems of life, then we can imagine this: there is no shit.

The ecosystems that sustain us may be terrestrial or aquatic and, because they are only loosely bounded (and perhaps just figments of our imagination), may overlap and straddle various borders. Species that live at the edges of water, such as hippos, bears, and otters, transfer nutrients across the water-land boundary.[2] This transfer across the boundaries does not negate the identity of the systemic boundaries, any more than eating and shitting negates your own identity as a person.

Fish can be important for seed dispersal of some terrestrial species. Some colleagues of mine, who work on mercury pollution in the Amazon, advised the people who live there to only eat fish that didn't eat other fish. This is great advice for protecting human health, since the top predators (fish that eat other fish) were the ones with the highest mercury concentration. However, some of the fish that don't eat other fish eat fruit that drops into the water, carry the seeds of the fruit trees to new locations, and poop them out. What, then, will happen to the fruit trees if people eat all the fish that transport the seeds? Who is going to do that ecological work on behalf of the fish? A generation from now, will there be fruit for the fish and the people to eat? So, what is good for individual animal species, such as ourselves, is not necessarily good for the ecosystems that will make future versions of ourselves possible.

The excreta of marine animals have important roles in aquatic food webs and ecosystems beyond seed dispersal. The feces of most aquatic animals, with a water content of over 90%, are held together in packets covered in a mucous membrane. Schools of fish follow hippos and manatees and feed on their droppings, or at least feed on the larvae that feed on the plankton (tiny, free-floating life forms) that feed on the hippo and manatee dung. Some

2 Sometimes this task is carried out at what we might imagine to be painful personal cost. One wildlife researcher commented to me that there are "so many shells" in otter scat, "you actually wonder what their anus is made of. Really, it must hurt terribly."

species of fish live dangerously and spend their nervous little lives nosing around the cloaca of whale sharks. There is nutrition to be had, and it is apparently worth the dangerous lifestyle.

In the beginning, then, were the single-celled organisms, floating in the warm sea, maintaining their membranes, sucking in chemicals and excreting waste, just stayin' alive. From that simple beginning, in many different places, in different ways, emerged celebratory tumults of interacting, co-dependent organisms, trading information, energy, and nutrients, and building complex relationships. Thousands of plant and animal species interact in such localized systems. Sometimes, these ecosystems and the organisms that make them up are based on a past that is suddenly unlike the future (during a shift from a warm Earth to an ice age, for instance). Then they die. What keeps going is the cycling of elements, information, and energy.

Having looked at evolving life as complex nested systems of organisms and multi-cellular plants and animals, we now take a radical shift in perspective. The plants and animals we see interacting on the landscapes around us can also be viewed systemically as comprising invisible cycling processes. Among the many possible elements and nutrients we could consider, three cycles in which excrement has important roles are worth describing briefly: nitrogen, phosphorus, and water.

Both nitrogen and phosphorus are vital components of amino acids, proteins, and genetic materials. Almost 80% of the atmosphere is nitrogen, but it is in a form that is not available to living systems. Certain kinds of specialized micro-organisms can take nitrogen from the air and "fix" it, that is, incorporate it into compounds that can be used by plants and animals. Some forms of cyanobacteria can do this, as can the *Rhizobium* bacteria that live in tiny nodules in the root systems of leguminous plants (beans, peas, clover, certain trees). This ability makes these bacteria essential for life, and it's why legumes are often included in crop rotations (clover being grown and then plowed into the soil to increase available nitrogen). Cyanobacteria present a more contradictory picture. Some cyanobacteria are used as nitrogen-rich fertilizer and food (for example, spirulina is considered by some people as a health food supplement in the U.S. and elsewhere). Other strains are associated with toxic "blooms" that can (and do) kill people and other animals. Once nitrogen is incorporated into biological communities, it can be recycled through various plants, animals, and bacteria, until it is put back out into the atmosphere through various forms of de-nitrification. Certain kinds of bacteria do this, often in areas where oxygen demand exceeds supply (such as in swamps or polluted rivers).

Phosphorus is an essential part of proteins and genetic material, of fats, cell membranes, and bones, teeth, and shells. Unlike nitrogen, most phosphorus is tied up in rocks and sediments, and leaches out into water, where it slowly heads to the ocean. The guano from fish-eating birds is an important mechanism for transferring phosphorus from the sea to the land, where it is reused by terrestrial plants and animals. Through shit and decay, a lot of this phosphorous returns to the soil, streams, and eventually to the ocean bottom and into the rock cycle.

Water is essential for life in all its forms, and must be reused for life to continue on the planet. Water is not actually any more scarce now than it ever was, and there is no more water now than there was at the beginning of the planet. It is only re-distributed, much of it by people. Some of this water is used directly by people; other water is taken up by domestic animals and plants. Eventually the water used by ourselves and our animals is excreted in shit and urine.

Powered by energy from the sun, water enters the atmosphere through evaporation and transpiration; later, powered by gravity, it falls to Earth (as rain or snow) or condenses (dew) and gets incorporated into bacteria, plants, and animals. Like most other land animals, people are on average almost 60% water (depending on fat content, age, and so on). Eventually the water runs back to the seas, which cover more than 70% of the Earth. The movement of animal feeds and human food around the world represents millions of metric tons of water. Those who are against selling water across borders might wish to include this in their calculations. I'll come back to this later.

Whether we view the ever-changing biosphere as a web of interacting organisms or as grand cycles of elements, none of what we currently understand as the Earth would have been possible without shit. From an ecological perspective, when we observe the production and management of manure, we should be thinking not only of contamination and health in relation to individuals, but also about the implications for seed dispersal; movement of water, elements, and nutrients; bacterial ecology; soil replenishment and impoverishment; and the long-term flourishing of life on Earth.

That is why the nature and distribution of animal feces is so important, and why global solutions to food production and manure management are often misguided and dangerous. Animal feces are essential ingredients that make places what they are, whether a forest glade, a waving grassland, or a deep mountain valley. Some animals, such as migratory birds, fish, and large grassland mammals (caribou, bison, wildebeest), have carried these nutrients and seeds, in their manure, over large distances, and thus enabled change and innovation in new places.[3]

What is fundamentally different in the twenty-first century is the way in which people have accelerated the pace, volume, and geographic extent of this movement of energy and nutrients. The world we live in is not just different than it was 100 years ago: it is profoundly, dramatically different. We cannot assume that any of the organizational strategies we have built over the past few millennia will enable us to adapt to what is coming tomorrow. We cannot address the issue of manure — indeed we cannot act morally — without somehow coming to terms with the speed, volume, and global reach of shit over-production from over-copulating fecund populations of people and the animals we have domesticated.

In order to imagine and act our way out of this Slough of Despond, it is important, but not sufficient, for us to begin to understand the general features of the ecological relationships that made us possible. By examining the cycles of how life-essential elements such as phosphorus, nitrogen, and carbon are cycled in ecosystems, agricultural and urban engineers can develop "best practices" for managing biosolids.

We need to go beyond this, however, to regain a sense

3 Rarely, live animals can also be transported to new, geographically dispersed living sites through feces. Japanese white-eye, or *mejiro*, is a fruit-eating bird well known to disperse seeds. In 2011, Japanese researchers reported that about 15% of tiny, 2.5-millimeter snails *(Tornatellides boeningi)* survived passage through the stomachs and intestines of the white-eyes. One of the snails even gave birth to young after emerging from the poop.

of home in the natural world that birthed us. Keep practicing that mind-shift from cells to animals to ecosystems; stretch even farther to go from sub-atomic particles to the universe. Then come back to where we live and think about how everything we observe works together to create who we are, and who we can become. One way to begin this re-imagining of ourselves is to consider the eating and shitting behaviors of individual animals, including people, in particular times and places. These behaviors, and the excrements associated with them, represent the "real world" mechanisms by which elements are recycled, landscapes are created, from which ecosystems emerge, and by which we have come to be who we are. This, then, will be the subject of my next few chapters.

CHAPTER 4

TURDS OF ENDEARMENT: WHAT EXCREMENT MEANS TO ANIMALS

Single-celled organisms depend for their lives on the membranes that surround them, and on the ability to draw in nourishment and to push out waste; ecosystems depend on co-evolved, geographically local rule-sets that characterize excrement and death and recycling. Animals are the most obvious, visible link that ties us all together.

Feces can be grossly classified by shape, color, texture, smell, and location in the environment, and each characteristic tells us something different about the life of the animal and its ecological roles. These attributes might be thought of as defining the public style of excrement — its presentation to the world, if you will — as differentiated from its substance. This is how we first notice excrement. We could also add taste, which is something some have explored as children, but there are not many people who would share such information publicly, so great is the wincing and tut-tutting in the audience when human feces–eating is described. For now, it is enough to say, it happens.

The nutrient content, scent, and structure of feces have

co-evolved to serve multiple roles within species and link a variety of plant and animal species into coherent ecosystems. Examining feces, by whatever method, can tell us a great deal.

Scent is a very important characteristic of feces, and not just because it annoys humans. In omnivores and herbivores, including people, the odors of feces can be attributed to bacterial action on undigested food and to the release of sulfur-containing molecules such as skatole, indole, mercaptans, and hydrogen sulfide. Many of the functions of dung within and between animal species — as differentiated from its ecological functions — have to do with the characteristic scents produced by the different chemical composition of foods eaten, how they are processed in the gut, and how they are expelled from the body. At least one Swedish researcher has argued that human excrement has much less scent if it is not mixed with urine, which provides nitrogen that results in some of the foul-smelling products.

Sharks can apparently smell human feces in the water from a mile away, depending on currents and the size of the deposit, but the scent of excrement is used by many mammals to better effect than attracting sharks. Carnivores, for instance, will sometimes eat herbivore scat to mask their own scent for hunting. Some pet dogs have retained this habit from the wild, and one can spend money to "cure" a dog of this behavior, although I would rather have a dog that eats feces than one that bites people. It may be that eating feces adds nutrients to the diet, or enables dogs to maintain a healthy bacterial flora through re-inoculation, or maybe it is just an evolutionary oddity, like singing in the bath to scare away predators (works for me).

Although many a veterinarian might think that the

anal glands of dogs were created to get infected, impacted, and squeezed by the nimble fingers of well-trained professionals who had to go to school for many years to learn how to do this, those glands serve other purposes as well. In carnivores (except hyenas), anal gland secretions, which adhere to the feces during defecation, have characteristic and complex odors that communicate such vital information as species, territory, sex, reproductive state, and geographic movements. This information is useful not only for other carnivores but for prey, to detect when a predator is near. Quite sensibly, cattle, sheep, and monkeys stay away from the smell of panther feces.

People generally do not use scent to track animals, but scent is still important in molding our reactions to it. Apart from the different scents given off by cattle, pig, and dog dung, certain diseases result in peculiar scents, because of the kinds of fermentation, carbohydrate or protein breakdown, or chemical production by particular microbes. One of my veterinary colleagues claims she can differentiate a coccidial from a parvoviral infection in dogs by smell, even from a distance. One wonders if feral dogs might use such scents to avoid diseased animals; doing so would certainly confer some evolutionary advantage — but perhaps the gain is outweighed by the strong selection pressures exerted by fast-moving road vehicles.

There is no doubt that the excretions of animals have always posed something of a challenge for some of the individuals and species who deposit them, and much of this has to do with scent. It is not always advantageous, for instance, to have your perfume announce your presence to other animals. This is particularly true for prey animals, where betrayal of one's presence has been dealt with by a variety of strategies such as coprophagy, burial, or putting the dung in places some distance from where they live. While the stories of birds carrying fecal sacs to drop into streams or sloths burying feces may be more common, some butterfly larvae have evolved more spectacular behaviors to lead predators astray, such as ballistic fecal ejection, sometimes called frass-shooting.

Brazilian skipper caterpillars, about 1.5 inches long, can shoot their feces six feet away, but most ballistic frass-shooters are satisfied with a 20- to 40-body-length fling — albeit at 1.5 meters per second. (I wonder, if at some future "Green" Olympic Games, humans might attempt to better those records.) In a variation of this, colonies of giant honeybees (*Apis dorsata*) — up to 40,000 bees at a time — have been known to fly en masse and drop about 20% of their body weight in yellow shit. During the Vietnam War, some military commentators, not knowing about the bees, wondered if this was some sort of biological warfare. Perhaps it was, but the war was being waged by another species, no doubt furious at human destruction of their habitat.

Not all animals have developed such dramatic tactics. Many, like humans, simply create a space away from where the food and babies are. This serves both to redirect the predators and to announce one's presence to friendly animals (perhaps of the same species). Pigs, for instance, do not naturally, as some have suggested, roll in their own poop. They have areas in the pens where they defecate, well away (if possible) from where the food is. They have this in common with llamas, who defecate in well-defined areas (a meter or two across); the impalas of Selous Park in Tanzania, who have their dung middens; giant South American otters, who trample a large communal toilet area along the riverbank to define territory; badgers; and African mole rats, who will sometimes set aside special dung-tunnels.

Two-toed sloths create dung middens to indicate mating sites not just for themselves, but for the moths that live in their coats. Genets are solitary, long-tailed, mongoose-like animals that create treetop or rock-top communal latrines. These serve as a kind of Facebook, allowing strangers passing through to sniff out potential mates or rivals. Bushbuck use defecation sites for intersexual communication. The females put out the signals and the males monitor the females for receptivity. One could speculate whether dung middens played any similar roles in human evolution, indicating places where other humans might be found for breeding purposes. Although "meet me tonight at the dung midden" seems to me to be a pick-up line with little chance of success, I confess that, in my mate-seeking days, I never actually tried it, so I have what is called an absence of evidence rather than evidence of absence. I would be interested to know if anyone has used it successfully.

The scents emanating from animal excrement are important, as well, for biologists who wish to study animal behavior and ecology. Examining the scats of animals (a so-called fecal-centric approach to studying wildlife ecology and conservation) is often as informative as, and more ethically defensible than, other more invasive methods of gathering information about the eating habits of wildlife, such as radio-collaring.

Finding those scats, however, is a big challenge, especially in wilderness landscapes. One tactic is to use dogs, who are great sniffers. They have already been used to detect drugs and other contraband at airports, and to track down crooks and hostages. Why not train them to find wild animal scats? In fact, this has been done. Karen DeMatteo, a biologist at the University of Missouri, has trained a Chesapeake

Bay Retriever named Train to detect scats of several South American wild animals. By studying the distribution of the hundreds of scats detected by Train, DeMatteo has been able to describe the habitat preferences of pumas, jaguars, spotted cats, and bush dogs. Another dog — a Labrador cross named Tucker which had failed as both a house pet and a police dog — was hired by the Center for Conservation Biology at the University of Washington. He is afraid of water but loves the scent of whale poop so much he will jump into the ocean after it, which is very helpful for those who wish to track whales.

A mixture of information from scat-sniffing dogs and the latest technologies for genetic analysis and DNA characterization can provide a wealth of data on where animals travel in the wild and what they eat. This is important evidence for determining the extent and types of habitats that need to be protected in order to prevent species extinctions and for understanding the relationships among multiple species on a landscape.

Once the scats are found, we can — like my guides in Selous — use the shape, color, size, and content of scats to identify which animals are around and what they have been eating. For instance, the bear scats I find on the Bruce Trail, along the Niagara Escarpment north of where I live, are usually thick logs chock-a-block with berry seeds. On the west coast of Canada, one might see fish bones mixed in with berry seeds in bear scat.

The characteristics of human excrement have been used to determine what people eat, and the health consequences of that diet, particularly in parts of the world where outdoor defecation is still common. Based on comparative studies between countries in Africa and Western industrialized nations, Denis Burkitt and his colleagues suggested that a fiber-rich diet helps prevent a variety of diseases and complaints in people including hiatus hernia, diabetes mellitus, coronary artery disease, colonic diverticulosis, colorectal cancer, appendicitis, varicose veins, and hemorrhoids. Burkitt also became famous for his numerous slides of human feces taken on his early morning walks in the bush in Africa. This, my wife should note, is a step beyond the pictures of animal shit that I have sometimes pursued on our holidays. He was apparently quoted as saying that the health of a country's people could be determined by the size of their stools — more fiber in the diet resulting in larger stools — and whether they (the feces, not the people) floated or sank. From these observations, North American baby boomers (myself among them, I confess) made a great obsessive leap of dietary faith and began a long romance with oat bran, wheat bran, and granola.

The land use and excrement consequences of this baby-boomer shift in diet have never been adequately explored. We are producing more excrement. But is it better excrement? Are the foods that end up in this excrement better for the soils? These questions are of far greater importance for the planet than whether or not we suffer from irregularity.

Burkitt's observations on fiber and health were almost single-handedly responsible for the huge demand in North America for high-fiber foods and diets. He made no reference to bear populations, however. Whatever the seeds in the scat say about the health of the bears, they do tell us that these animals are important in the ecology of the plants in the region, carrying seeds to new places, along with the remains of fish, which contain some excellent fertilizer to get the plants off to a healthy start. Take away the bears, and what happens to the berries? What happens to the riparian zones along streams frequented by bears?

While bears are key species in the ecological systems they call home, other animals have played a central role in the development of human cultures. Herbivores are capable of digesting the cellulose of plants and converting it into milk, blood, meat, and other materials, which can then be digested by people. It is perhaps no accident, then, that the woman on the airplane asked about sheep, cattle, and horses. All three, it turns out, are well adapted for life on semi-arid grasslands and a nomadic life, which represent the origins of modern human societies.

Sheep, like all their even-toed hoofed relations (even-toed ungulates, also called artodactyls) except for cattle and buffaloes, drop small piles of cylindrical or rounded pellets usually pointed at one end and concave at the other.

Cattle, buffalo, and bison leave behind flattened feces that accumulate in circular piles, which we used to call pies or pats. These are sufficiently well defined that they attract flying insects, which lay eggs, which turn into larvae, and in so doing turn shit into edible protein for birds. They also attract dung beetles, which can turn the dung into another generation of dung beetles.

Horse droppings, which look similar to those of wart-hogs — a random piece of information to share with your seat-mate in the airplane — are said to be kidney shaped, but to me they look like dark rye buns. I am not alone in this perception, as at least one enterprising company has developed what they call "bun bags" to catch horse manure before it hits the road. Someone must have once imagined that horse turds look like apples; hence the term "road apples," although that American slang term was apparently first applied to traveling actors.

Why do these differences in shape and appearance exist?

Cattle and sheep are both ruminants, and both are

grazers; that is, they eat mostly grass, rather than shrubbery or tree leaves.[1] The rumen part of their multiple-stomach system is like a great big fermenting and digesting barrel whose functions I described earlier; you can hear the rumbling and gurgling sounds (called "borborygmus") if you place your ear against the warm fuzzy upper left flank of a cow, just ahead of where the hip bone sticks out. This rumen soup has, to some of us, a rich and sweetly sour scent. Digestion is helped further when the ruminant belches up some of the food, re-chews it, and then re-swallows, which is called rumination (and is only vaguely related to what your parents did when you first asked them about sex).

There is nothing more wonderful and peaceful than sitting on a straw bale in a barn with a herd of ruminating cows. I hope I can die in those surroundings, or am at least reincarnated there, which is a more likely possibility.

Even though sheep and cows both graze, sheep get much of their water needs from the grass itself, and so drink a lot less water than cattle; their feces are drier and more discrete. Sheep are thus more efficient in water use, which is why you see them in desert environments more often than cows. They also eat closer to the ground, which is why they have been used to try to decontaminate radioactive pastures (by eating and removing the radioactive plants). This feeding behavior also has larger ecological implications. Australia ships about 25,000 tons of sheep meat and more than a million live sheep to Saudi Arabia every year: what, one might ask, are the implications of this for the movement of nutrients out of Australian soils

1 In aquatic systems, grazers are animals such as snails and beetles that eat algae.

and into the Arabian peninsula? Of course, sheep do sometimes graze on wet, verdant pastures, and there produce what biologist Ralph Lewin calls morular "shitlets," a term which I find endearing and shall have to find greater opportunity to use.

Giraffes, like goats, are primarily browsers, which has nothing to do with their use of the internet; they prefer foliage from trees and shrubs, rather than low-growing grasses. They are also ruminants and are nature's way of transferring nutrients from high up in the trees to the bacteria, protozoa, and dung beetles on the ground. They are very efficient in water use, and can go longer than a camel without drinking. Hence their feces tend to be similar to, but larger than, sheep pellets, like chocolate-colored baseballs.

Horses don't have the complex stomachs of ruminants. They do, however, have a big cecum, a large sac where the small intestine joins the large intestine, like an enlarged appendix. Because of this cecum, they are sometimes called "hindgut fermenters." The cecum is full of billions of bacteria and protozoa that help break down and ferment plant fiber. The horse gets some benefit from this as fluids and some nutrients can be absorbed in the large intestine. However, since most of the nutrients are absorbed into the body from the small intestine, much of the work of these microbes in the cecum is for the benefit of the ecosystem at large rather than for the individual horse. The horse has to compensate for this relative loss by eating a greater volume of hay or grass, or wait for indulgent owners to feed them oats, which are more nutrient-dense than hay or grass. One might wonder if this process led to fertilization of grassland where horses (and the cows and sheep and their owners) lived, enhancing the growth of the grasses, and hence helped the population sustain itself, even if it was of no great benefit to the individual horse. Some evolutionary biologists would consider this heresy, whatever the evidence, but I tend to prefer to refer back to evidence, rather than ideology, in trying to understand the world.

In any case, turds of horses are firmer than those of cattle because they contain a lot of undigested straw and chaff, and they aren't slurried up in rumen juice.

For those of you readers who wish only to impress your neighbor in the airplane, you may stop reading here, and respond to her. For the others, who are, I hope, most readers, and who are unexpectedly interested in the subject of excrement and its importance for ecological sustainability, please read on. Maybe on the next plane ride, you can be the one to ask the difficult question to put off an annoying seat-mate, or engage someone with the same warped turn of mind as you (and me).

Rabbits, hares, and pikas are also hindgut fermenters, but they get around the nutrient loss problem by eating some of their own feces. Like horses, they drop hard, fibrous shitlets, but unlike horses, they also squeeze out clusters of soft, dark, sticky currant-like pellets (cecotrophs). These delicacies are formed in the cecum, which in the rabbit is ten times larger than the stomach. Made up of fermented food and bacteria, and coated with mucous, cecotrophs are passed at night in domestic rabbits (which eat and pass normal stools during the day) and during the day in wild rabbits (which are nocturnal). They are usually eaten directly from the anus, so that they do not dry out or lose their nutritional value.

Watching a rabbit carefully nip cecotrophs from his anus must be an educational experience, surely, for pet rabbit-owning parents and children alike. "Daddy, what is Bunnykins doing?" asks the child, trying to figure out if he could manage the same dexterous act, the anxious parents fearing that, limber little kid that he is, he just might be able to. Perhaps this will lead to a career in yoga instruction, the parents never once suspecting the origins of such a choice in a family that only ever played hockey.

For non-humans, coprophagy — the deliberate eating of shit — is a behavior that has evolved in situations where survival trumps moral pretention. While most people know about rabbits, dedicated scientists (or sometimes just people with time on their hands and no access to television) have observed various excrement-eating behaviors in horses, rabbits, capybaras, ringtail possums, guinea pigs, mountain beavers, lemmings, chinchillas, nutrias (coypus), rats, mice, gerbils, degus (bush-tailed rats), and other rodents, dogs, cats, and shrews.

Eating shit has evolved among non-humans to fulfill both nutritional and protective roles. Parents want to prevent predators from detecting the scent, particularly of newborn animals. A deer doe, for instance, will eat her fawn's feces for the first month of its life in order to avoid attracting predators. Funny, I don't remember that from *Bambi*. Some songbirds also eat the feces of their young (also don't remember that from Disney), especially when the nestlings are very young, a practice that some behavioral ecologists have attributed to good bird economics: the parents are avoiding the transportation costs of fecal sac disposal. Still, many species of birds do pick up the mucous-coated packages of baby poop and drop them into streams or ponds nearby, especially as nestlings get older. Which means that if the birds disappear (for instance, through deforestation or predation from the roaming of gangs of unemployed feral cats), the richness of life in the waterways is also diminished.

In several species coprophagy is about fostering health and protecting against disease. Rabbits are gaining proteins and water-soluble vitamins. Mice are said to gain vitamin B12 and folic acid from eating feces, and if you prevent lab rats from eating their own feces, they don't grow well and can develop vitamin B12 and vitamin K deficiencies.

If a cow is sick and not eating for a long time, the microorganisms in her rumen will start to die off; one way to get her eating again is to siphon the rumen contents from a healthy cow (just like siphoning gas from a car!) and infuse them into a sick cow. Like community motivators and facilitators, the incoming bacteria will kick-start the moribund population of bacteria and protozoans and get them living and working (and having fun?) again.

Termites, too, consume their own excrement and undigested debris (frass), in order to acquire necessary gut flora, including protozoa, to aid in digestion.

A variation of this idea, called competitive exclusion, involves getting beneficial bacteria into the gut to keep out bacteria one doesn't wish to have there. The basic idea is simple: if there are beneficial bacterial already occupying a certain ecological niche, then the pathogenic bacteria cannot get a foothold. In Finland, where competitive exclusion was first developed as a poultry disease management technique, newborn poults (baby turkeys) were fed cocktails of mature turkey feces in order to prevent *Salmonella* infection.

People, apart from children and those with disordered mental states or abnormal appetites (a state called "pica"), do not normally eat feces, although there may be some titillation associated with eating something that has been associated with, or derived from, shit. In 2011 Mitsuyuki

Ikeda, a Japanese researcher at the Environmental Assessment Center in Okayama, reported that he had fabricated artificial meat from sewage containing excrement. The "meat," which is said to taste like beef, is 63% proteins, 25% carbohydrates, and 3% fats. Apparently much of the food value comes from the bacteria in the feces.

This risqué skirting of the impolite may explain at least part of the fascination with "Kopi Luwak," the name given to coffee made from beans that have passed through the bowels of Asian palm civets. The civets eat coffee berries, the coverings are digested, and the seeds are pooped out. It has become famous, or infamous, especially as a result of Jack Nicholson's love for it in the movie *The Bucket List*.

The coffee's infamy comes from its expense, from the ambiguities about its quality, from its association with feces, and from periodic rumors that it may be a hoax played by marketing-savvy Indonesians and Vietnamese coffee sellers on wealthy Westerners. Others may drink it to honor the peasant workers who discovered it: because the Dutch colonialists prohibited their Indonesian workers from picking coffee beans for their own use, the workers scavenged, cleaned, and roasted seeds from Asian palm civets' droppings. Because the taste of any Kopi Luwak depends on the coffee beans that were eaten by the civets, the flavors vary widely. The Balinese variety I am drinking as I write this, a gift from an Indonesian colleague, has none of the sharp, rich scents and pleasantly bitter aftertaste that I usually associate with coffee. The chocolate-colored powder has a faint, musty, earthy scent, as does the smooth drink prepared from it. Massimo Marcone, a food scientist at the University of Guelph, has studied the properties of the beans, and found them to be lower in protein (apparently a source of bitterness in coffee beans) and to have different volatile compounds, than standard Colombian coffee beans. These differences from "ordinary" beans are the result of enzymatic action in the civet's stomach.[2]

Despite the cringing responses of most people to the very idea of coprophagy, the principle of competitive exclusion used to such good effect in turkeys provides the basis for an effective therapy for a deadly hospital-acquired infection by *Clostridium difficile*. These organisms, part of a group of anaerobic organisms that are widespread in nature, occur at low levels in many animal intestines, including those of people. In many North American hospitals, *C. difficile* has been transformed into a killer bug as the result of intensive antibiotic treatment, which has killed off competing, non-killer bacteria. An important part of the treatment involves taking bacterial flora from the intestines (that is, the shit) of healthy donors, which are then infused through enemas or nasogastric tubes into the patient. This replenishes their healthy gut flora, which then out-compete the disease-causing *Clostridia*.

Competitive exclusion may also explain the fact that, in some studies, dogs that eat excrement are *less* likely to be found infected with certain bacterial pathogens such as *Clostridium difficile*. Having said this, I should point out that poop-eating dogs are *more* likely to shed some other dangerous bacteria such as *Salmonella*. The fact is

2 What is often missed in the cute culinary story of Kopi Luwak is the importance of the civet behavior in natural systems. In a natural system, the poop helps the wild civet to mark territory; the ecological effects of this would serve to disperse the coffee plants to new locales. What people view as a culinary curiosity was thus, evolutionarily, important for the survival of the civets and coffee plant populations in the wild. Today, most Kopi Luwak is produced by farmed civets.

that bacterial ecology in shit is not something we understand very well, and we should manipulate it with care.

Hyraxes are cute rabbit-like animals that live in trees and rocks in eastern and southern Africa; the rock hyraxes peer at you as you walk through their land, like creatures from a Star Wars movie. The tree hyraxes make a sound in the night like a creaking door, followed by a blood-curdling scream, which is quite unnerving if one is sleeping in a tent in what one thought was the middle of nowhere. Hyraxes create dung middens, and urinate on them to increase the scent, as territorial markers. These animals are related to Sirenians (dugongs and manatees), elephants, and aardvarks, and have multi-chambered stomachs in which bacteria break down plant fibers, their abilities in this regard being similar to ungulates. The Sirenians poop in water, which means that most of us are rarely afforded the opportunity to see the shit's shape.

Elephants are, of course, considerably larger than hyraxes, which explains why their daily output of big coarse cylinders can in no way be referred to as shitlets. Elephants can both browse on shrubs and woody plants and graze on grasses. When they browse, they can bring down whole trees; if populations are crowded together, this can be very destructive. However, elephants only digest about 40% of what they consume. Thus they are bringing down nutrients from high places and making them available for animals that are more constrained in their reaching ability, but less finicky in their eating habits, and are willing to nose around in elephant shit.

The 35-million-year-old aardvarks suck and lick and dig termites out of the ground, a diet which they supplement with a kind of underground tuber (*Cucumis humifructus*) whose seeds they disperse when they bury their feces; this burying is good for the ecosystem, but not so good for those who keep a journal of fecal sightings.

Medieval English hunters referred to the feces of hunted animals as "fewmets," a term derived from Old English *feawa*, meaning "scant" or "few," and "encounter" (*metan*), which suggests that wildlife turds are difficult to find. The turds of aardvarks seem to be fewmets indeed. Some use the term "fumets" to refer to deer droppings. The difficulty in finding feces from many wild species (and hence the need for zoologists to train sniffer dogs) reflects the conflicting implications of announcing one's presence in the woods, and the various behaviors that have evolved to deal with that conflict. It also reflects the fact that any organic material is, given the right temperature and moisture, quickly used as food by bacteria and fungi, broken down into materials that are usable by other animals and plants. Usually, if you find a pile of excrement (or a dead bird) it is either very fresh or there is a lot of excrement (or a lot of dead birds) around.

What emerges from these observations of animal shitting behavior is that a variety of species has survived because the types of food they can process, the types of feces they produce, and the behaviors that evolved to dispose of those feces are not only of benefit to themselves, but to the sustainability of the ecosystems within which they live.

Animals do not plan to enrich the ecosystems in which they live. Nevertheless, the co-evolutionary, complex webs of eating, defecation, and renewal across species have provided habitats that enable those animals to thrive. Feces, which are food for bacteria, fungi, plants, and micro-fauna, are "reborn" in the form of new food and shelter for the descendants of the original animals.

Ecologically, defecation behaviors of all species are a kind of gift giving that binds us together in a beautiful community of life, birth, eating, shitting, death, and rebirth. When we eat, we are taking a gift from the biosphere. When we shit, we are giving back. Our eating and defecating behavior says far more than any voting behavior about what kinds of citizens we are on this planet. This is why, fundamentally, shit matters.

CHAPTER 5

MAPQUEST TO DIARRHEA: THE FECAL-ORAL ROUTE

In spring 2011, a mutant, severely pathogenic, and antibiotic-resistant strain of *E. coli* spread across thirteen countries in Europe, sickening more than 3,000 people and killing forty-eight. The normal home for all *E. coli* species, most of which are law-abiding, contributing members of society, is in the intestinal tracts of warm-blooded animals — that is, in excrement. This epidemic, however, was spread through fresh sprouts from an organic farm in Germany. The original contamination source was identified as fenugreek seeds from Egypt. The genetic make-up of the strain of *E. coli* includes material last seen in sub-Saharan Africa.

When I first started teaching about the epidemiology of foodborne diseases in the late 1980s, such epidemics were sufficiently rare that I could peruse the literature every once in a while and come up with a few good teaching examples. By the time I retired from teaching in 2011, we were dealing with a global epidemic of epidemics, and I could barely keep up with the daily reports coming from all directions.

In the twenty-first century, epidemics and pandemics of fecal-associated infectious diseases have become relatively routine in industrialized agrifood systems.[1] And, although the origins of the bacteria involved are animal feces, they are now often spread through foods from plants.

For the billions of city dwellers and peri-urban "country-lovers" on the planet, and for the government leaders who respond to them, excrement is most often viewed as a threat to public health (especially for city dwellers) or a smelly environmental nuisance to neighbors of hobby farms. The health risks associated with feces have to do with direct toxicity, with the local spread of contaminants and infectious disease agents, and with widespread contamination of the food system. Although most discussions of the health risks of excrement focus on the spread of infectious diseases, it is worth at least mentioning problems of direct toxicity, which are mostly hidden from public view.

In the late 1970s, when I was in veterinary college, one of our clinicians was called out to a dairy farm to look at a "downer" cow. Dairy cows fall down and are unable to get up for a variety of reasons, often having to do with metabolic imbalances or calving problems. The clinician loaded up several senior veterinary students and headed out to see what was the matter. When they arrived, they could see the cow lying down in the barn, but the farmer was nowhere to be seen. The vet and two of the students went into the barn, bent down to check the farmer, who was lying next to the cow, and immediately fell to the floor. The third student, seeing what had happened, telephoned for help.

1 For more information on foodborne diseases, see my book *Food, Sex, and Salmonella: Why Our Food Is Making Us Sick* (Vancouver: Greystone Press, 2008).

The dairy barn had a slatted floor, below which there was a manure pit, where the cow dung, mixed with urine and water, was collected for loading into a "honey-wagon" and then spread on crop fields. The manure slurry in such pits is periodically agitated, releasing a variety of gases. Some of them are merely stinky, and others toxic. The most toxic of them, hydrogen sulfide (the rotten egg-smell gas), causes irritation and dizziness at lower levels, but effects can escalate to respiratory collapse and death pretty quickly. Making sure all the fans are working to keep the building over and around the pit ventilated is critical. Farmers and farm workers who forget this can be killed by manure gases. The farmer, the vet, and the vet students were taken to hospital and all survived.[2] These sorts of incidents happen every year on livestock farms; not everyone survives. While it is not a huge problem in terms of numbers, it is a tragedy for each affected family.

For non-farmers in the countryside, the downside of applying manure to the fields is the smell nuisance, rather than a problem of direct toxicity. But there are more serious, often hidden, issues. Nitrates from manure, like those from manufactured fertilizers, have been a serious problem almost everywhere that farmers have reared livestock under intensive conditions. The most disturbing effects of these nitrates getting into water systems are arguably ecological — the eutrophication of lakes and the growth of algae blooms that can be toxic and can change the ecology of water bodies in fundamental ways, most of which, such as the recurring "dead zones" of Lake Erie, do not promote human

2 Sadly, I do not recall the fate of the cow. Reported cattle deaths from hydrogen sulfide are rare; the gas is heavier than air, and only affects animals whose heads are close to the floor, or when the manure pit is agitated, in which case animals are often outside the barn.

happiness. However, there are also worries about human health. Exposure to high levels of nitrates, which the liver converts to nitrites, have been associated with methemoglobinemia, which results in lowered capacity of the blood to carry sufficient oxygen, as well as some forms of cancer. In the Netherlands, in the 1990s, 65% of drinking water came from groundwater wells, at least a quarter of which were seriously contaminated with nitrates. Several European governments implemented policies to restrict livestock rearing based on manure output. Although many livestock producers complied, some larger industrial livestock-producing companies elected to move elsewhere, where the regulations were less restrictive.

The general wisdom now, in many industrialized countries, is that farms should not raise more animals (pigs in particular) than they have land that can safely absorb the feces. This depends on the volume of manure, its texture and nutrient content, and the type of soil to which it is applied. Global warming and unstable climatic conditions, which alter the ecological movement of various bacterial and chemical components of manure, complicate this. The general principle, while well intentioned, is difficult to interpret in specific cases. Some European regulations, for instance, lumped cows and other bovines into the same category. Subsequently Italian researchers determined that the manure of buffalo, used for producing mozzarella cheese in Italy, is lower in nitrogen than cow manure and should be managed differently. The contradictory challenge is that countries and trading partners want general rules that apply to everyone, but the ecological interactions between animals, manure, landscapes, and ecosystems are always local and historically conditioned.

The direct impacts of excrement on farmers' health,

water potability, and soil structure rarely make it onto the radar of most urban consumers. This means that politicians tend to ignore the issues until a crisis arises. Of greater concern to most voters is the possible contribution of manure to the spread of disease. This is understandable, because the effects are immediate and, for individuals at least, gut-wrenching. So, while city-based politicians may minimize direct environmental issues related to manure, babies with diarrhea are another matter entirely.

Diarrhea, which clinically is when your excrement takes the shape of the container into which it is put, leads to dehydration and death. Death, especially of babies, is heart-breaking.

Globally, more than a third of the world's population (about 2.5 billion) lacks access to potable drinking water and water for washing and toilet facilities. About 2 million people die every year from diarrhea; most of these are children. According to reports from the World Health Organization and UNICEF (United Nations Children's Fund) about 1.5 million children under five die from diarrhea every year — more than from malaria, measles, and AIDS combined. Diarrhea is almost always caused by fecal contamination of food and water. Simply put, disease from fecal contamination occurs when people and other animals poop upstream from where people take their drinking water, and the filtration plants that have been put into place are unable to filter all of it out. In the twenty-first century, everybody is upstream from someone else.

Although the human disease outbreaks can be serious and disturbing, the underlying problems that give rise to them are rarely related to health systems or medical care delivery. The interactions between ecological webs, land use, and social dynamics are often more important, and are not often recognized. It is therefore on these interactions and their effects on people and other animals, rather than on medical descriptions of disease, that I wish to focus this discussion.

An outbreak in Canada in 2000 that sickened half of the 5,000 people in Walkerton, Ontario, and killed seven provided a tragic example. The outbreak could be attributed to a complex systemic failure, involving higher than normal rainfall (probably related to global warming), the presence of livestock in the watershed, the soil types and the way the land sloped, a badly placed well, government downsizing and unregulated decentralization, lack of education for the water managers, poor communication, and incompetence. The overriding lesson from this tragedy is that, in the face of increasing and irreversible global climate and economic instability, attention to ecology and local adaptive capacity is critical.

Many instances of fecal contamination of the local environment are more stealthy than the Walkertown outbreak. In the 1980s, while I was in the middle of my fieldwork for my Ph.D. in epidemiology, there was an outbreak of *Salmonella* on some Ontario dairy farms in the area where I was working. In that outbreak, the organism spread through feces within and between local farms. There was some evidence that the bacteria got into surface water and drained down to a nearby lake, used by people for swimming. Studies done in Canada in the 1990s showed links between manure-spreading on fields, the occurrence of pathogenic forms of *E. coli* in local wells, and higher levels of *E. coli*-related diseases in rural people as compared to city people. Pathogenic strains of these bacteria, which can cause severe kidney damage and death, have been found in run-off water in areas with high cattle density.

Antibiotic-resistant strains of bacteria, along with the antibiotics themselves, can be found in manure run-off from farms where those antibiotics are used. Researchers have found tetracyclines, tylosin, sulfamethazine, monensin, nicarbazine, and amprolium in swine, cow, and turkey manures. Furthermore, antibiotics absorbed into soils have been found to retain their antibacterial properties. Neither the public health nor the ecological impacts are known, but I suspect that the ecological impacts on shifting bacterial ecologies in the soil may be the more serious.

The presence of disease-causing bacteria from animal excrement in the food system is related to the mixed blessings of livestock in agriculture. Animals are an important functional element in healthy agroecosystems (processing inedible plant material to nutrients useful for other animals) and an excellent source of food and income for people. *And* they are a source of disease. William McNeil (*Plagues and Peoples*), Jared Diamond (*Guns, Germs, and Steel*) and Ton McMichael (*Planetary Overload*) have made convincing arguments for the emergence of infectious diseases in human societies as the result of settled agriculture and domestication of livestock some 10,000 years ago. Certainly almost all the infectious diseases of importance in human history have come from animal populations, or, like tuberculosis, went from people to animals and then back again. Many of these diseases have had decisive impacts on the rise and fall of empires and the overall course of human history.[3]

Livestock, including the foods they eat and the manure they produce, perform a variety of important ecological

3 For more on diseases that are shared by people and animals, see my book *The Chickens Fight Back: Pandemic Panics and Deadly Diseases That Jump from Animals to Humans* (Vancouver: Greystone Press, 2007).

and social functions on farms. How did they become a problem, and why, over the past 10,000 years, did many of the great livestock-associated diseases disappear after they had caused such great damage? Why do they then re-appear in certain circumstances?

An epidemic depends on what is called the "probability of adequate contact" between infected people or animals, and susceptible people or animals. Once a certain proportion of the population (the general rule is about 70%) has gotten the infection and either recovered with immunity or died, the epidemic dies away. At that point, the chance of an animal or person who is carrying the agent coming into contact with a susceptible person or animal becomes vanishingly small. [4]

This is referred to as herd immunity: the population is immune, even though many individuals are not. If there are a lot of vaccinated or otherwise immune individuals, the sick people or animals and the "pre-sick" ones are less likely to meet each other. Those who have not been ill, or who are not vaccinated, are protected by those who have been vaccinated. Some uncharitable souls would call this freeloading.

After an epidemic, the agent may linger, without causing any problems, in a few individuals (a sub-clinical infection, or carrier state). It causes sporadic cases of disease when those people encounter susceptible individuals or a new epidemic when they encounter whole susceptible populations. When Europeans sailed to the Americas, the agents they carried in their bodies found susceptible individuals and left millions of dead aboriginal people in

4 Thus, after sweeping from the Far East across Central Asia and into Europe, killing millions, many major infectious diseases such as plague and typhus died out.

their wake. The great pandemics that destroyed the First Nations in the Americas have died down, defeated not by modern medicine, hospitals, and vaccines (which had their greatest impacts after these epidemics had mostly disappeared), but by herd immunity, as well as better nutrition, public hygiene, and housing.

Survival patterns in my own familial history can be at least partly attributed to herd immunity developed by continual exposure to animal manure over many generations. At my maternal grandparents' home in the Ukraine — in a practice dating back to the sixteenth and seventeenth centuries — the barn was attached to the house, and there were two bedrooms in the barn itself. Cow dung was sometimes used as fuel both in Europe and later in North America in the form of dried pats, as it is in much of South Asia today, or it was molded into bricks. In at least one family, so I have been told by a relative, the mother would spread cow dung on the dirt floor, perhaps as a sealant or insect repellent. Animal-related disease never seemed to be a problem — or at least was not reported to be a problem — and the benefits of living with livestock far outweighed the risks. During the Russian Civil War and famine, my maternal grandparents had ready access to milk without having to wander out into dangerous public places. But more importantly, my teenaged mother and her sister were able to escape through the attached barn when the house was attacked by bandits, hiding in the hedges and fields around the homestead until dawn. I wouldn't be here if some sanitary inspectors had insisted that the barn be set so many meters away from the house.

So why, in the latter part of the twentieth and early twenty-first century, have new animal-related infectious diseases emerged, and old ones re-emerged, in our agri-food systems? Through economies of scale, encroachment on new habitats, and global trade, we have created new pathways between sources of infection and previously un-exposed populations. We have, in the language of epidemiology, increased the probability of adequate contact between the disease reservoirs and susceptible people. The cattle in the barn and the people at our dinner table are no longer on the same premises. They are rarely even in the same country. Yet, through the agri-food system, we are linked more closely than ever, eating off the same plates and bathing in the same waters.

Many of those agents with global pandemic status — *Salmonella*, pathogenic *E. coli*, and *Campylobacter* being the most prominent — are of animal origin and spread through manure. Influenza viruses have their origins as enteric (fecal) viruses in wild waterfowl. In this case the threat of an avian influenza pandemic is caused by changing probabilities of adequate contact between wildlife, farm animals, and people, through incursion of farms into wetlands, and international trade.

Exploding populations of people and domestic animals, combined with rapid, worldwide urbanization (centralization of human shit production), rapid growth in economies of scale in agriculture (centralization of domestic animal shit production), and global trade and travel (widespread redistribution of shit in all forms) provide for multiple pathways from the bum to the mouth. In other words, we have created new pathways for the pathogens in excrement, thereby increasing the probability of adequate contact.

The very long list of excrement-related diseases that have plagued people, sickening and killing millions, includes cholera, hepatitis A, salmonellosis, campylobacteriosis, diseases caused by *E. coli* O157:H7, giardiasis, amoebiasis, and, more generally, travelers' diarrheas going by various names such as Montezuma's revenge, Casablanca crud, and Delhi Belly. Global pandemics of cholera, the most recent of which killed tens of thousands of people in the Americas in the 1990s, and, since 2000, epidemics of shit-derived *Salmonella*, *E. coli*, and *Campylobacter* in a variety of vegetable crops illustrate the widespread fecal contamination of the food system.

The simplest way to frame the public health problem of excrement in the agri-food system is to say that there is human and other animal shit in the food and drinking water of billions of people and this sickens or kills them. The solution to this would appear to be more and better waste management technology, better flush toilets, sinks to wash hands, and filtration plants. Such solutions worked in the nineteenth and twentieth centuries: flushing shit down the toilet and then setting up a treatment plant to filter out the lumps from the water before we drink it was clearly better for human health than throwing shit out the window. These actions no doubt saved millions of human lives and may be, from one vantage point, one of the greatest public health interventions in history.

But getting shit out of the house is only part of the issue, and, if we are not careful, the kinds of technology we use may make things worse, given current world populations of people and other domestic animals. We may simply be pumping manure from one house to another. Or we may be solving part of the problem (disposal of excrement) and creating a much more serious one by drawing down the water tables and magnifying drought. We can live — albeit uncomfortably, distastefully, and often dangerously — with shit. We can't live without water. In a century when everything is unstable — climate, economy, politics, human behavior — building bigger plants and better toilets will not be enough to resolve the issue.

When excrement production and processing is centralized, bigger plants lead to bigger failures, and there are always failures. When excrement is increased in amount but the animals are dispersed, as in urban dogs and cats or park-loving Canada geese, it isn't channeled into a single sewer pipe anyway. The leakage and spread of feces across the landscape, or "non-point-source pollution," is a kind of guerilla shit-warfare by Mother Nature, and our best technologies are helpless. From an ecosystemic point of view, excrement shows its wicked self quite clearly.

Two parasites of public health importance can help us reflect more deeply on the geographic scope of the twenty-first century manure problem and its relationship to ecological processes.

Toxoplasma gondii is a tiny parasite that lives, and sexually reproduces, in the intestines of cats. Felines are the only type of animal that can shed cysts in the feces, and then only when, as kittens, they get infected the first time. In kittens, the parasite may cause depression and loss of appetite. There is also evidence that *Toxoplasma* infection in animals can alter behavior. Infected rats are less afraid of cats, and hence are more likely to get eaten, closing the infection cycle.

It is as a disease of people that toxoplasmosis is of greater concern, however. In most adults, it shows up as fever and aches and sometimes as tiny cysts in the eye (floaters). Some researchers have suggested that *Toxoplasma* cysts in the brain are related to schizophrenia. If a woman gets infected for the first time when she is pregnant, she can miscarry, or have a stillbirth, or her child can later have learning disabilities. Once someone is infected, the cysts hide in the muscles and organs, where they generally don't cause problems unless the person is immuno-suppressed, at which time the cysts "wake up." *Toxoplasma* encephalitis from such revived cysts was a significant cause of death in the early days of the AIDS epidemic.

Public health workers used to be concerned that pregnant women and children might be exposed to *Toxoplasma* by cleaning out litter boxes or playing in sandboxes. In the past few decades, this concern has taken on global dimensions.

An outbreak in 1995, in Victoria, Canada, was a kind of warning flag. The Victoria outbreak, with more than 100 confirmed clinical cases and an estimated total number of infected people ranging from 2,895 to 7,118, was the world's largest up to that date. The source was apparently cats (domestic or wild) defecating in the city's reservoir. Once researchers started looking beyond the sandbox, they began to find *Toxoplasma* everywhere.

We have evidence that muskoxen, caribou, bears, sheep, and wolves across the north and west coasts of North America have been exposed to *Toxoplasma*. It has even been found in sea otters and harbor seals. So how are they getting cat shit? The current assumption is that the parasites are spreading from cat feces that are flushed into urban sewage systems, or from the direct defecation of the millions of feral, or just free-roaming, house cats into watersheds, and hence into the ocean. Imagine a world with lots of feral cats and then increasing dry spells followed by heavy rains (flushing). Don't imagine long. That is the world in which we live.

A second parasite, *Giardia*, enables us to visualize in other ways the ecological webs through which excrement travels. Giardiasis, caused by a cute single-celled "animalcule" first seen (in his diarrheic stool) by Anton Van Loewenhoek through his primitive microscope in 1671, is another great poop-traveling parasite. Anton must have been amazed, as the parasites appear for all the world like ghostly, stringy-haired heads, complete with eyes (their nuclei) and a mouth (something called a median body), swimming by. I know it is just my perception, but I swear they look as if they are smiling in a smug, self-satisfied way. And well they might.

The disease caused by *Giardia* is not inconsequential, being associated with diarrhea, gas, gut pain, and loss of appetite. And treatments are not without their own discomforts. Erin Fraser, a veterinarian who is the managing director at Veterinarians without Borders/Vétérinaires sans Frontières - Canada, studied backyard chickens in Honduras. When I visited her in the remote village where she lived, she looked as if she was just going to waste away completely. Giardiasis, she guessed. And she had heard stories that the treatment was also rough, so figured she would "ride it out," since, in many healthy people, giardiasis can be "self-limiting" — an interesting turn of phrase used by veterinarians and physicians to describe diseases that disappear on their own, after a while. And yet, aren't all diseases self-limiting? Indeed, aren't we all self-limiting? As Erin's academic advisor, I confess to having been more than a little worried. But she survived.

It has been called "beaver fever," because wild beaver are known to sometimes carry it — as are dogs, cats, cattle, and children. The term "beaver fever" was originally coined when hikers got sick drinking from a clear mountain stream — presumably contaminated with beaver feces. Most people get it from their daycare-aged kids, who don't always wash their hands after they poop. Daycare centers are also significant sources for adults to get hepatitis A, another fecal-oral disease, increasingly common in North America. Because the effects of hep A are caused by a functional immune system attacking infected cells, and children don't have fully functional immune systems, the hepatitis A virus doesn't always bother the kids, but when they bring it home, Mom and Dad can really get sick.

But to get back to *Giardia*. Beyond backpackers and daycare, we often think of it as a disease of poverty or at least poor sanitation, which it is. The Crusaders in the Middle East apparently suffered from it, a message from God that many Muslims and Jews might have wished had been heeded. This little parasite has been around a lot longer than people; indeed, some scientists say (and some dispute) that the ghostly smilers have their origins back at the emergence of eukaryotic life itself, a couple of billion years ago. Enjoying as they do a watery habitat, they have certainly found plenty of places to call home. And once warm-blooded animals evolved, they quickly found the perfect love-boat — mammalian feces — within which to multiply and hitch a ride to see the world.

Giardia is the most commonly identified intestinal parasite of humans in Canada and the United States; in developed countries worldwide, the rate of infection is about 2% in adults and 6–8% in children. In countries with poorly developed water sanitation infrastructure, about a third of all people come down with the infection at some point in their lives.

There really seems to be no escaping *Giardia*, and the parasite tells us a great deal about the ecology of feces from warm-blooded animals. Recent studies of *Giardia* demonstrate that widespread infection is not restricted to humans, and indicate that encounters with infected excrement are pretty much inescapable. Maëlle Gouix, a French wildlife field researcher working in Canada, wrote to me:

> My colleague and I worked in very remote places in Canada, unspoiled, almost pristine old growth forest. We had to travel every day, by foot, no roads, no hiking trails, sometimes only wildlife trails . . . a paradise. . . . We were looking for wolves and wolf scat. We did find a lot of both. Back in the lab . . . we checked out scat for parasite eggs, and found a lot of them, and compared them to dogs from communities, to see if those dogs might introduce parasites into native wolf populations. [. . .]
>
> We found *Giardia* in both wolves and dogs. But the more parasitized ones were . . . us. After a few days spent in one of the villages there, my co-worker and I started experiencing the same growing uncomfortable symptoms that weeks spent side by side allowed us to share. One day she called me in the lab and said, "Come and check this out!" The slide was massively covered with thousands and thousands of fluorescent *Giardia* (we used immunofluorescence for diagnosis). "Wow, cooool, we've never had such a parasitized one before, where is that from?!" She laughed and said it was from her. Because the parasite is so widespread, we are not really sure where we picked up the infection.
>
> I don't know if you've had it before, and people react differently anyway, but holy cow, those nasty little buggers are terrible, I had no clue!
>
> The muskoxen in the North have also got *Giardia* from people, but how? It's such a big area and so

> little populated. . . . And when they slaughter wild muskoxen, the whole thing (400 muskoxen killed sometimes) happens right by the village. And musk-oxen guts and waste are left on the ice by the shore. *Giardia* [cysts, being resistant to harsh environments] . . . get in the marine ecosystem and maybe contaminate men again via marine mammals or fish or clams.

The excitement of the researcher is palpable; it is the excitement of making a discovery that is unexpected and unwanted. Nevertheless, there it is: from the arctic to the tropics, from modern urban centers to small rural villages, the fecal-loving parasites are thriving pretty much everywhere.

Finding *Giardia* and *Toxoplasma* in unusual places is important for understanding the parasites themselves. However these findings are equally important for understanding the nature of the challenges we are facing in the twenty-first century. The world is pretty much splattered with excrement. Another sewage treatment plant is unlikely to solve this.

With outbreaks and epidemics of fecal-related viruses and bacteria such as *Salmonella* and *E. coli* being reported in many wealthy, industrialized countries reported to have the "safest" food supplies in the world, and being associated with spinach, almonds, bean sprouts, tomatoes, and other organisms not known to have intestines, the announcement that excrement is everywhere is not news.

Bacteria and viruses can be destroyed by most water treatments, or by cooking food (if you are open to cooking your salad), but the little parasites are another matter. In the external environment, parasites such as *Toxoplasma* and *Giardia* are transformed into cysts that are resistant to many of the usual weapons we use against living contaminants in water. UV treatment appears to be one of the few effective technologies against *Giardia*.

To round out this discussion of shit-related human disease, I should add that not all excrement of health importance is from birds or mammals. Even insect frass is important.

"Kissing bugs" (Reduviid bugs) are vectors for Chagas disease, sometimes called American Sleeping Sickness. At night, these bugs creep out of cracks in walls or the ceiling and down the ropes of hammocks in the shantytowns of Central and South America. After injecting a small amount of anaesthetic, they suck blood from the medial canthus of the sleeper's eye. When finished, they take a crap. The sleeper awakes, his eye itches, and he rubs the frass into his eye, hence injecting the parasite, *Trypanosoma cruzi*, into his bloodstream. Many years — often decades — later, about a third of people so infected develop flabby heart or bowel muscles and die a long, lingering death. The infection is for life, so although concerted efforts to control the bugs from the 1980s on reduced the annual number of new human cases in Latin America from more than 700,000 to about 40,000, there are still 10 to 15 million people infected. Effective responses require substantial political and economic commitments, including not just spraying for bugs, but also investments in better housing, economic and social equity, managing land use, and education.

Although much attention on manure-related disease focuses on human health, there is also a significant, and often complex, impact on managed wildlife, where diseases' effects are mediated through animal behavior, human behavior, and ecological processes.

Jan Myburgh is a wildlife toxicologist and ecohealth investigator who has done a lot of work in Kruger National Park in South Africa. Kruger Park is oriented from north to south (for the convenience of people, to free up agricultural lands to the west of it) but traversed by rivers that go from west to east. Although there is a fence on the west side of the park, farmers draw as much water out of the rivers as possible before the watercourses flow into the park. Some nutrient runoff from these farms also seeps into the rivers and hence flows into the park. As water levels drop during the dry season, the park managers create earthen dams on the rivers to make small water holes to ensure enough water for the animals.

Many animals, frustrated that they cannot follow their natural east-west migration routes because of both the design of the park and the human populations settled along its boundaries, crowd at the water holes and shit.

Hippos, in particular, spend most of their lives in water, staying cool and moist in their preferred habitat. Except for marking their trails away from the water at night, hippos usually defecate in the water. As the dry season sucks the water out of the already-polluted rivers and shrivels the lakes, hippos crowd into the water holes and defecate there. Since there are fewer water holes, the density of hippos and concentration of hippo shit increases in those that remain.

As Myburgh tells it, the hippo dung in the water results in blooms of cyanobacteria, organisms that can often outcompete others for nitrogen and phosphorus. The cyanobacteria in these water holes are toxic to many species. Elephants plunge joyfully into the water, splashing and playing and dispersing the toxins. Buffalo crowd in too, the front ones getting pushed in by the back ones, which also scatters the blooms. Rhinos, however, stop at the edge of the water and lean daintily forward, sipping — not wishing to get their toes wet — and get killed by the concentrated toxins.

If they die, stressed-out, at the water hole, their infected carcasses and feces can wreak further havoc by leaking disease-causing organisms into the water, adding insult to the cyanobaterial injury. Antelope also stand at the water's edge, with the wind in their face so they can smell predators coming. Unfortunately, they are standing in the place toward which the algae are being pushed, and they also get intoxicated and die.

The story of the hippos, rhinos, and water management strategies in Kruger National Park illustrates the unintended consequences of well-meaning but simple-minded interventions in a world where behavior and ecology interact in unexpected and complex ways. Building parks to conserve animals, managing water so that they have enough to drink, or farming to produce food can — if we pay insufficient attention to the ecological and cultural context — unintentionally affect animal populations that are ostensibly being protected. The threats to animal well-being in this case came not from direct attacks, but from engaging in "ordinary" behavior in extraordinary circumstances.

Similarly, the most serious threats to human well-being usually come in a guise and from a direction least expected. We — scientists especially — are good at focusing, but, as clinical neurologist Oliver Sacks has argued, we do not sufficiently honor our peripheral vision.

Those concerned with excrement-associated public health and wildlife problems usually focus on the negative impacts of high concentrations of manure in and around areas where there are high concentrations of people and domestic animals. What I wish to do in this book is to enable us to cultivate our ability to pay attention to what is happening beyond the immediate, to enable us to re-imagine both the problems and the possibilities for a way out of this mess.

CHAPTER 6

HERCULES AND ALL THAT CRAPPER

The ways that other animals have met the challenges of managing their feces are, in many instances, reflected in the ways that people have dealt with shit. After all is said and done, we are animals and share a great deal with all other animals — emotion, morality, devious behavior, genetics, problems with shit. We are also a very particular kind of animal, with our own cultural histories and hang-ups.

The way that human cultures have traditionally dealt with shit thus reflects a mixture of our biological inheritance and the cultural rituals we have created to reinforce or redirect various behaviors. We are conflicted: our instincts are to protect our young (and hence to remove feces), to mark territory (keep the feces close by), improve food supplies (use the feces for fertilizer to grow food), and prevent disease (keep the feces away from food).

Our language reflects our conflicted attitudes — not just with regard to what we call the stuff itself, but what we call the act of defecation, and the places where we defecate. As anyone who has traveled becomes aware, not only is talking about the act of wanting to take a dump itself problematic, but where one moves one's bowels is also complicated. Does a person really rest in the restroom, hang water in the water closet, or wash in a latrine (which comes from the French *lavare*, to wash)? And if, as Lewin asserts, the term "covering one's feet" in the King James version of the Bible refers to the act of defecation, is this because the desert tribes of the Middle East did not have access to water when they went number two? And if a dog lives in a doghouse, what lives in an outhouse? And going to the john? I thought this was something that sex-trade workers (as they are properly called today) did, but not the rest of us; apparently, however, we all do it regularly, at least if we do not suffer from irregularity. Or is "going to the john" simply a public relations ploy by the Jakes of the world, whose name was the medieval English term for a toilet, to clear their name and dump the ordure on Johns? What about a privy? Does this have any connection to what really goes on in the privacy of the Privy Council, that advisory gang to the government in a parliamentary system? Some birders talk about their birds "venting," vent being the term sometimes used for the cloaca, that joint exit hall for feces and urine found in birds, reptiles, amphibians, and marsupials. Remember this the next time someone complains to you about their day and says they are "just venting."

If we take a long view (more than the trivial few millennia that many in our culture concern themselves with), we might say that the ways in which human societies have managed their own — and their animals' — dung can be understood not only as emerging from cultural habits and taboos. They also reflect evolutionary inheritances from other species and ecological patterns of nutrient recycling. The risk of talking about evolutionary precedents of human fecal-related behavior is that we might be tempted to create a false sense of uni-directional evolutionary development. But history is not always a story of progress, or even linear change.

Many of us have been well schooled in linear narratives of human progress, and the assumption might be that there is some kind of uplifting, progressive story to be told about how humans have dealt with excrement over the millennia. Alas, it is not so. The narrative of shit is one of fits and starts, water and earth, discovery and loss and rediscovery. What has changed is that there are so many more of us now, and more of our domestic animals, than ever before in history. The narrative is one of us trying this, that, and the other thing, and then trying the first thing again while the dung piles up around us.

Various human societies use a variety of strategies to deal with excrement. Some developed earlier in our history, some later, and we can learn from all of them. There never was, nor is there now, a "perfect" one-size-fits-all solution. In ecology and evolution, the diverse interplay of context with content, nature and nurture, genetics and social-ecological landscapes is everything.

One way in which humans have dealt with excrement is associated with nomadic peoples who, having left poop behind in the woods, later returned to those same areas to discover that their feces had fed the next crop of nuts and berries. The co-evolved lifestyles of people, horses, cattle, and sheep in semi-arid landscapes reflect this integration of culture and nature.

It was only with the beginnings of settled agriculture in the Neolithic era (about 10,000 BCE and continuing to the present) that humans have had to face the problem of having to remove excrement from us, rather than us from it.

As people settled and created farms rather than wandering through the bush, they began to create large accumulations of solid waste in small areas. This led, in various times and places, to the creation of special rooms inside houses, special pots, outdoor latrines, and areas in the woods that were designated for defecation. Unlike those species that have used toilet areas to meet and greet, we have tended to separate the places for feces from other living spaces.

In some cases, human waste was buried. The buriers included several First Nations groups in the Americas, as well as the Essenes, a mystical and ascetic Jewish group that lived in desert areas and in cities throughout what was called Roman Judea, from about the second century BCE to the first century CE. The Essenes also buried their libraries, which may or may not be significant in regard to the relative values they placed on scrolls and poop. The scrolls, which fed their minds, later came to be called the Dead Sea Scrolls. The poop may have fertilized the fig and olive trees that provided their sustenance. The parallel with burying feces is thus not a frivolous one, as the burial of the texts allowed the preservation of ideas and stories, and their renewal centuries later, in much the same way as the feces renewed the soil.

People, like many species, developed alternatives to midden heaps to meet mates. This being removed from the equation, the major remaining advantage to leaving animal and human manure in particular places was that the plants would grow better in those places. There is evidence that Neolithic settlements disposed of their excrement in and around their settlements, probably in dung heaps. Manure from livestock has long been recognized by farmers as a source of fertilizer and returned to the soil, and seems, until recently, to have posed less of a problem than human excrement. This distinction may reflect the observations of our ancestors that exposure to human feces was more dangerous to people, in terms of disease spread, than exposure to the feces of other animals. In simplistic evolutionary-selection terms, children who were exposed to human feces were more likely to die (from cholera and other fecal-borne diseases) before they could reproduce, than children not so exposed.

Never content just to gather in small groups, our forebears began to move into cities, where crowding and large-scale human feces production generated new challenges. When populations are small, a latrine hole for a family or a dung heap (like an animal midden) just outside a village or farm are effective ways of moving the excrement away from human habitation. What does one do in a city? Rooms in houses where one would just take a dump on the floor were fine if you had extra rooms in which to store the excrement, or long-suffering servants to clean it up, and as long as the dung holes were not directly connected to fans circulating air in the rooms below (hence giving us the expression "when the shit hits the fan"). Throwing poop and urine out the window — along with the shout "*Agua!*" or something similar to warn passersby — was an effective way to get it out of the house in parts of medieval Europe. Still, one would have to walk through it on the way to the opera. Very annoying.

While farmers have often used manure as fertilizer, the use of human manure in this way is less consistent. The Chinese, with one of the world's most intensive and, until recently, most sustainable agricultural systems, have a history of collecting and marketing human excrement (called "night soil") that goes back 3,000 to 4,000 years. Researchers estimated that, historically, 90% of all human excreta in China was being recycled in this way and that it provided about a third of all the fertilizer used in that country.

The Japanese too have a long and disciplined use of human excreta in agriculture that predates cities such as Edo (now Tokyo) but that flourished especially as the country urbanized. Farmers provided buckets beside the fields and asked travelers to use them when they defecated. In a web of transactions that mimicked natural cycles, the seventeenth-century city of Edo sent boatloads of vegetables and other farm produce to Osaka to be exchanged for the city's human excrement. As the cities and markets grew (Edo had a million people by 1721) and as intensive paddy-farming increased, prices of fertilizers, including night soil, rose dramatically; by the mid-eighteenth century, the shit owners wanted silver — not just vegetables — for payment.

So who owned this shit? Apparently the people who owned the buildings were the proud proprietors of the excrement produced by the tenants. If some occupants within an apartment left, the owners increased rents for the remaining occupants, since they had less excrement to offset the basic capital costs of running an apartment block. The prices for human excrement were so high in eighteenth-century Japan that stealing human shit was an acknowledged crime, punishable by imprisonment. In a pattern not unfamiliar to contemporary farmers caught in a squeeze of high oil prices, the cost of shit became so high that poor farmers couldn't afford it. Although the practice of using human excrement for fertilizer declined after the fall of the Tokugawa regime in the late nineteenth century, and during the industrialization of Japan in the twentieth century, one researcher has estimated that at least 50% of human waste in Japan is still treated and used as fertilizer.

The use of human and/or other animal feces for fertilizer has been and remains a worldwide phenomenon, a kind of convergent cultural evolution. Although this recycling has been more prominent in the Far East, the use of manure as fertilizer has been discovered and embraced in just about every settled society. In the Aztec city of Tenochtitlán (now Mexico City), excrement and organic waste were gathered and sold for fertilizing crops or tanning hides. In Peru, the Incas stored, dried, and pulverized excrement for use on maize crops. Ibn al-Awam, an Arab living in twelfth-century Spain, described composting techniques incorporating human excreta; the use of this compost as fertilizer for plants was said to cure illnesses in banana, apple, peach, citrus, and fig trees as well as in grape vines, palm and cedar trees, and wheat plants.

In the Middle Ages, as people moved to the cities, and cities produced excrement, Europeans commonly used excreta and gray water (water that has been used for washing) on gardens. The Cistercians, near Milan, used refuse, excreta, and wastewater on their land starting in about 1150. The people of Freiburg, Germany, irrigated their meadows with excrementally enriched wastewater from at least 1220. The irrigation and cultivation of these meadows improved growth in dry periods, was said to reduce the incidence of plant pests, and contributed to stabilizing the nutrient balance in the meadows. After reaching a peak in the nineteenth century, these irrigation practices declined and were mostly stopped by the 1960s.

By the nineteenth century, along with rapid industrialization and urbanization, the use of raw (untreated) sewage for fertilizer became widespread in Europe and the United States. New Yorkers made a profit selling their manure to surrounding counties as fertilizer (and then importing it back in the form of vegetables).

In the twentieth century, the rapid, uncontrolled growth of urban populations, which continues globally to this day, the development of chemical fertilizers, and an understanding that fecally contaminated water could be a disease threat to humans shifted the arguments to favor public health outcomes over recycling. Farmers and veterinarians might well tolerate the scent of manure, as it signals a useful product that can replenish the soil, stimulate crop growth, and improve profits; urban dwellers are more likely to associate the scent of shit with filth, disease, and the endless work of farming from which their parents escaped.

In summary, taking the long view, we can see that our conflicted cultural and personal attitudes toward shit have deep roots. In evolutionary terms, positive associations with the scent of excrement may be rooted in biological urges to define territory and communicate with others (as described in Chapter 4), as well as the observation that food plants grow better in areas that had been manured. When human populations were mostly nomadic and when settlements were small and sparse, the positive associations with excrement outweighed whatever risks were perceived. This positive view of excrement has persisted when connections between city and countryside have been explicit and open (as in the story of Edo), and in rural agricultural areas today. As we have increased our understanding of transmission of killer diseases such as cholera and childhood diarrhea in the last few centuries, however, and as the beneficial association of flush toilets and clean bathrooms with health has become clear, city dwellers have learned to take an unambiguously negative attitude toward shit. The shift from a positive view to a negative view is thus rooted in shifts from people living in the country to people living in cities, to a loss of connection between food producers and consumers, and to our increased scientific understanding of causes of disease. But the connections between our attitudes to shit and the subcultures we inhabit are even more complex.[1]

The notion of perfume reflects this subcultural complexity. How one interprets a scent — whether one thinks of it as fine and lovely, or filthy and disgusting — is culturally conditioned. One allegory from the sixteenth century describes how the perfume of a courtesan was apparently thought objectionable by angels, but that of an "honest" dung-collector's cart was considered just fine, and even virtuous. Which, if nothing else, says something about the cultural preferences of angels that inhabited Europe in those years. At that time, civet musk — the buttery scent from the perineal glands of civets — was an ingredient in heavy perfumes used to mask the odors of unwashed upper-class bodies. In the seventeenth century, William Shakespeare's son-in-law John Hall was in good medical company when he used it to treat hysteria in a woman by rubbing it on her navel. It is not clear if the treatment was effective. Although synthetic alternatives were available

1 And often very personal. When we had babies at home, I used to complain of the scent of infant feces while changing diapers, without much sympathy from my wife. She could not understand how I could tolerate, even enjoy, the odor of cow manure but not the smell of my own children's poo. I suspect it is an evolutionary thing, my wanting to keep that child smell away from the house so the predators don't come by, and liking that of cows because it means food. Being the great male provider, it is important to know where to find the cows. At least that's my story, and I'm sticking to it.

by the 1940s, the pheromone derived from civet musk has been used as a stabilizer in perfumes until relatively recently.

In some sperm whales, the accumulation of unregurgitated indigestible materials mixes with feces and intestinal secretions to form a solid, fragrant substance called ambergris. Ambergris, highly prized by the perfume industry, may be expelled into the ocean, or may kill the animal by obstructing the intestine. Christopher Kemp, writing in *New Scientist*, asserts that the "rich, complex odor" of ambergris "has been compared to fine tobacco, the wood of old churches, the smell of the tide, sandalwood, fresh earth, and seaweed in the sun." Musk and ambergris, both associated with excrement, but put to good use in "high culture," may serve to help us think in new ways about manure (or perhaps to re-think what we mean by "high culture").

In eighteenth- and nineteenth-century Europe, bodily odors were still associated with "lower" classes, and cleanliness was next to godliness, so that the use of perfumes (still fortified by civet excretions) and powders displayed moral as well as economic superiority. This morally superior attitude persists throughout the industrialized world today, along with the general mythology that the poor people living in slums actually like to be without proper toilet facilities, or at least can tolerate this situation better than rich (white) folks with more sensitive dispositions. In every culture, from ancient Rome (where the main sewer of the city, the Cloaca Maxima, was cleaned by prisoners of war) to eighteenth-century England (where cesspool cleaners were required to work at night), those who manage feces on behalf of the rest of us are among the least regarded workers.

Today, urbanites in industrialized countries who move into the countryside admire the cows — and complain about their smells. In this, they reflect the urban-agricultural split I described earlier. Of course, there is a qualitative difference, and not just in terms of odor, between the happy scents of a few cows or pigs rolling in the straw and the flood of concentrated effluent that exudes from a modern industrial livestock facility. Humans seem able to pick up the scent of pig manure from a great distance. This has led to millions of dollars, including $1.7 million in President Obama's 2009 stimulus package earmarked for Iowa, being poured into research to make pig shit smell nice.

Perhaps our aversion to pig manure in particular is related less to the manure itself than to an ambiguous relationship with the animals it comes from. *Trichinella spiralis*, which is associated with muscle aches and pains and occasionally death, is acquired by eating infected pork. Pigs pick up the parasite by eating rats or, sometimes, each other. *Taenia solium*, which is at one stage in its life cycle a tapeworm in human intestines, can cause epileptic-like seizures. In this case, humans acquire the intestinal infection from cysts in pork, and pigs get the cysts from eating human manure. People get the brain cysts (and the seizures) when they ingest the tapeworm larvae from their own feces on unwashed hands or contaminated foods.[2]

2 As I have often told my food safety students who want to eagerly lecture the public about what to eat and what to avoid, our eating habits are not entirely formed around disease risks. Despite the risks, people eat whole mussels, oysters, and clams, including the entire digestive tract full of excrement; being filter feeders also makes these shellfish an excellent source of whatever viruses or bacteria are in the water they are taken from.

We cannot undo our evolutionary and cultural histories. The global population is now mostly urban, mostly amnesic about the ecological and economic benefits of manure, and increasingly aware of the disease problems associated with shit. Some will argue that the public health problems we face are not entirely new, and not particularly wicked, although they are more acute. These people, many of them my colleagues, and indeed myself not too many years ago, will argue that we have the tools to fix the problem, that they have been tried, and that they work. The flush toilet, considered by some to be the greatest public health invention ever, is one of those tools.

Water-flushed latrines, which emerged as the dominant manner of waste disposal in the twentieth century, have ancient roots. The fifth labor of Hercules required him to clean out, in a day, three decades of accumulated dung from about 3,000 cattle in the stables of the Greek King Augeas. The manure pile had not seemed to be a problem for the beasts themselves, but, according to the great twentieth-century scholar, novelist, poet, and reteller of Greek myths Robert Graves, "it spread pestilence across the whole Peloponnese."

This was probably the first recognition that manure runoff from intensive feedlots and other intensive animal-rearing operations could be a source of disease for others. Hercules accomplished his feat by breaking two holes in the walls around the cattle and then diverting the Alpheus and Peneus rivers through the cattle-yard, thus effectively flushing all the shit away.

Hercules had been promised a tenth of the cattle as his reward, but after Copreus (which means "dung man"), the herald, told Augeus how Hercules completed the task, the king refused to give him the cattle, arguing that the rivers, not Hercules, had achieved the desired cleaning. (I like it that the messenger in this story was called Dung Man, as it fits in neatly with my argument that the shit is telling us something.) Thus the magic of flushing, the idea of rivers as sewers, and dilution as the solution to pollution have long been entrenched in the consciousness of human civilization.

Flush toilets were available in many houses in the ancient cities in the Indus Valley between 2500 and 1500 BCE, among the Romans a thousand years later, and, at least by the eighteenth century, in northern parts of Europe. Much later than Hercules, and thinking on a scale less heroic, one might refer to Sir John Harington in the late sixteenth century, who wrote instructions for how to construct a "valve closet," and Alexander Cummings, who patented a flushing device in 1775. The term "water closet" only entered the English vocabulary in the nineteenth century to distinguish it from the "earth closet," a popular composting toilet developed by Reverend Henry Moule in 1860. It was basically a seat over a bucket, and pulling a handle dumped earth or ashes on the excrement. But the bucket had to be emptied somewhere, which created other problems. In a densely populated city, where would one empty it?

All of which is a circuitous path to assert that Thomas Crapper, the man who popularized the flush toilet in the late nineteenth to early twentieth century, might have been a clever capitalist but hardly original in his thinking. In technology as in science, the claim to novel ideas is often made and is usually fraudulent. As an aside, the *bidet*, or "pony" in French, which was ridden to clean one's butt, not to flush away excrement, never really caught on with the English. It is more akin to the notion of having a small bucket for scooping water from a water-filled concrete tank, as is found in some Asian countries. The water is used to wash one's privates (with your left hand, please!).

While a flush toilet improved the hygiene in a home, what to do with the results of the flush has been a challenge from the start. In some societies, when it wasn't thrown out the window, people emptied their waste into backyard cesspools. In sixteenth-century France a 1539 edict *required* people to build their own cesspools. One author, Dominique Laporte, has argued that European programs for shit management were based on a bourgeois social construct related to the primacy of individual property rights.

The Romans saw the disposal of excrement as a public, rather than a private, issue. One of the earliest sewage systems, Rome's Cloaca Maxima, built by Tarquinius Priscus (616–578 BCE), was originally designed as a system of channels to drain rainwater. Only later did it develop into the city's main sewer, transporting up to 45,000 kilograms of excrement a day, discharging it downstream from the city into the River Tiber.

The sewers of industrial Europe were originally designed to carry away wastewater and kitchen slop but not excreta — although city dwellers were known to raise their skirts and skirt the laws on that. When piped water was introduced (improving cleanliness in the home), the cesspools overflowed, creating stink and disease in the neighborhoods.

By the early nineteenth century, problems with backyard cesspools were sufficiently serious that city dwellers in Paris and London were officially allowed to connect their cesspools to the city sewers. It was seen as an advance over throwing shit out the window or letting it pool in backyards and leak into the streets. But city sewers were not designed for this volume of human feces. Who would have thought we could produce so much shit?

As any epidemiologist knows, what is good for an individual is not necessarily good for the population. Flush toilets and connecting to city sewers merely scaled up the issue of too much shit from household to city, and the problem needed to be solved again at that level. Emptying cesspools into the sewers and the sewers into the rivers was, on the one hand, a colossal waste of nutrients. More dramatically, it was also associated with cholera outbreaks. In the first half of the nineteenth century, tens of thousands of people died in Paris and London from cholera.

By the mid-1850s, when the Thames River flowing through London became known as the Great Stink and its odors threatened to overwhelm legislators, England passed laws that helped transform the city's sewer system. In 1849, reportedly more than 250,000 cubic meters of sludge were discharged from sewers into the Thames, and at least one company providing drinking water had its intake pipe only a few feet from a leaking cesspit in which a baby's nappies had been washed. The baby was sick with, and later died of, cholera.

In 1854, Dr. John Snow (who also introduced Queen Victoria to anesthesia during childbirth) demonstrated, through investigating and mapping cases of cholera in Soho, London, that the disease was being spread by water contaminated with human feces. In a controversial move, Snow removed the handle from the Broad Street public water pump, which, according to his maps, was a major culprit. Snow has since been celebrated as the "Father of Epidemiology." His astute and accurate observation and documentation of disease patterns, and his inference of the vehicles by which diseases were spreading, were remarkable because he had no knowledge of the actual disease agent itself. This was, after all, well before bacteria had been identified.

Some, less kind, might attribute greatness to the man as an epidemiologist because he recognized that the epidemic had peaked and would decline no matter what he did, a trick known and loved by other epidemiologists seeking public approval since that time. In fact, Snow himself is said to have recognized this as a possibility, which nevertheless did not change his general conclusion that the disease was being spread through fecally contaminated water.

Still other critics, the more cynical ones perhaps, point out that government officials scoffed at Snow's explanations and replaced the pump handle after the epidemic subsided. They were offended at the wild theory that the disease was spread through fecal-oral transmission, and perhaps irked that an upstanding and wealthy water company was having its name besmirched in such an unprovoked manner. In any case, the idea that shit in drinking water might not be a good thing eventually took hold, as the evidence connecting contaminated water with the disease accumulated. Sanitation engineers and scientists joined forces with (or themselves became) politically active reformers. Finally, the "germ theory" of disease being put forward by Robert Koch and Louis Pasteur, which substantiated what the sanitary engineers were saying, replaced the old "miasma theory" that asserted that disease was generally caused by bad air.

The upside of this new germ theory and its champions was that it led to sewer systems, water systems, and flush toilets that much improved the health of urban populations in Europe. The downside was that the discovery of bacteria reinforced the idea that feces were *only* inherently dangerous and dirty. The truth is that human and animal feces, if not handled correctly, and if they are allowed to contaminate food and water, often are dangerous. This is particularly true of human feces, since they are the most likely to contain microbes that are adapted to living in people. These bacteria may include those associated with cholera, or typhoid fever (caused by a human-adapted *Salmonella*), or *Clostridium difficile*, the current bane of hospitals, which lives in at least a small percentage of apparently normal human intestines. Yet, if handled in such a way that the dangers are mitigated (for instance, through composting, or bio-gas production), feces of all sorts, including human feces, can be immensely beneficial. It is this latter truth which seems to have been lost in the campaigns to promote public health in the past century.

And so I circle again back to the wicked culture-nature entanglements with which excrement confronts us. On the one hand, our practical farming selves know that animal manure is an excellent fertilizer, enabling human settlements and people to thrive and sing and write great literature. Thus, almost every society has groups within it that manage excrement as a technical issue whether it be as fertilizer on farms in Europe or fuel in Indian villages.

On the other hand, there has been a general cultural tendency almost everywhere — probably related to urbanization and crowding, probably rooted in legitimate fears of disease — that there is something filthy, dirty, and embarrassing about taking a shit, or even talking about taking a shit, or "talking about not talking about it" (as 1960s guru-psychiatrist R.D. Laing would have said).

Even anthropologists and ethnographers in the European and North American traditions, who have been almost obsessive in describing the various eating and sexual behaviors of "other" cultures, have tended to shy away from considering what medical anthropologist L.L. Jervis calls "exuviae and exudations." This is not so much shame (although showing one's naked backside to others certainly can involve some of this), as it is, I think, embarrassment at being an animal. I recall, at the age of nineteen, being doubled over with diarrhea in Calcutta and having to drop my pants in a small alley off a main street in that bustling city. In some ways, as a callow Mennonite boy who had been taught that belief systems and spiritual well-being were all that mattered in the world, it was an epiphany. For all our lofty philosophies and religious visions, we still have to shit.

The most difficult temptation to resist when addressing issues that evoke conflicting emotions in us, is to make generalizations based on personal experience and anecdote. Some authors have argued that the link between evil and shit is universal, pointing to feces as an accessory to witchcraft for Indonesians, Patagonians, the Dyaks of Borneo, and the aborigines of Victoria. These authors have also invoked Gnena, a demon believed to live in feces by the Bambara (in Mali) and a Korean evil spirit that frequents toilet sheds. The Thai Filth Ghost went beyond frightening people while they were shitting to sometimes reach into their anus and pull down part of the rectal lining. While now the butt of jokes (according to my Thai friends), the Filth Ghost tale might, in earlier times, have accounted for hemorrhoids.

These stories reflect a mixture of genuine fear, nervousness about slightly embarrassing behavior, and sometimes humor. In this context, it is no surprise that belief systems, religious and otherwise, have had a strong influence on how we think about and manage feces. Acts of urination and defecation have often been linked to ritual impurities. In some cases, arguments framed as sacred prohibitions in traditional cultures have modern scientific justification. Whereas a Thai villager might not defecate in a moving stream because it offends Mother Water, a Canadian biologist would discourage this activity because it contaminates drinking water and can spread diseases such as cholera, hepatitis, or giardiasis.

Sigmund Freud and other psychoanalytic investigators over the past century have deemed it important to describe human behavior related to feces in terms of characterizing personal development and personality types. Hence, in popular culture, we now encounter the idea of "anal-retentive" personalities — those perfectionist, obsessive, neurotic, keeping-it-all-in people many of us know (and, to some extent, all are) — and "anal expulsive" personalities. One might also explore the relationship between fear and voiding behavior, and our somewhat nervous responses to stories of people being, literally, scared shitless. The first reported defecation in the face of terror is in the Annals of Sennacherib in 701 BCE. When the kings of Babylon and Elam fled in terror from battling the Assyrian Emperor Sennacherib, it was said that they "let their dung go in their chariots." Better in a chariot, I suppose, that in the shark-infested waters off the Australian coast.

Reflecting on our psychologically ambiguous relationship to excrement is useful for therapy, perhaps. But for purposes of constructing the larger ecological and evolutionary narrative of our species and its place on this planet, these ways of framing the issues seem less relevant. We will make more progress if we look at how some cultures and religions have recognized the dangers inherent in feces even as they found ways to take advantage of its useful properties.

Islamic societies have characterized feces as being a source of *najassah*, or impurity, requiring a ritualized wash after exposure. Nevertheless, as I understand it, composting transforms feces from something impure to something useful. This, it seems to me, is a reasonable response, since the heat generated by proper composting kills most pathogens. In Indonesian Java, which is at least nominally Islamic, human excreta are added to fish ponds and hence transformed into useful food through dilution, flow, and consumption by species less fastidious than ourselves. Whatever the rationale, such practices promote both human and ecosystem health, so that they can be justified on both scientific and religious grounds, a situation devoutly to be wished for on many other issues.

Adding composted feces to the earth is a good thing, and is generally acknowledged to be so, provided that feces are properly composted, and not just buried. The crucial step of composting is reaching high enough temperatures to kill the pathogens and weed seeds. That we all defecate and thus can make a daily personal contribution to the renewal of the planet would seem to be something to be celebrated. In the last century, the ascendancy of fossil fuels — coinciding with the ascendancy of industrially manufactured fertilizers, the population explosion, and the flush toilet — has taken a source of celebration and made it into a problem to be solved.

How can we recover, if not a sense of celebration, at least a sense of meaning, in our shit-making? Where could our culture look for inspiration? I am reminded, once again, of the dung beetles. My favorite name, among all the dung beetles, is the genus of Sisyphus. In Greek mythology, King Sisyphus, an avaricious, lying, violent, clever man, who promoted navigation and commerce (an ancient counterpart, perhaps, to neo-liberals who promote global trade in order to enrich themselves), thought he was smarter than all the gods. After one trick too many, the gods punished him by having him roll a giant boulder up a hill. Just before he reached the top, the rock rolled back down and he had to start over again. Forever. In "The Myth of Sisyphus," an essay written in 1942, in the midst of World War II, Albert Camus saw in this myth life's fundamental absurdity. Yet Camus imagined that Sisyphus could be happy in the midst of his meaningless toil, and concluded that the "struggle itself towards the heights is enough to fill a man's heart. One must imagine Sisyphus happy."

But the scarab gives us a more hopeful — I would say a more realistic — spin on this myth. The rolling of the giant dung-ball up the hill is not meaningless at all. It embodies the deepest meaning of life, a re-imagining, in biological terms, of what John Donne meant when he said that we should not ask for whom the bell tolled — it tolled for all of us. The scarab exemplifies for us the reinvention of life from its own waste, the expectation of a future re-awakening, the celebration of the dance of all living things, even through the most mundane of daily activities. Every day, in everything we do, we are renewing and reinventing life on this planet. In identifying with the dung beetle, Sisyphus — humanity itself — is redeemed.

CHAPTER 7

THE OTHER DARK MATTER

The unwelcome presence of excrement in food and water and piled up in various places on the landscape is usually addressed as an "environmental problem." The framing of an issue as an environmental problem assumes that there is a source that pollutes, and a localized landscape that is contaminated. There is rarely any recognition of far-reaching unintended consequences for plant, animal, and microbial species through impacts on food webs and ecological cycles. The focus is on a well-defined problem (parasite pollution in a river) and its cause (inputs into the river from poor placement of a sewer outlet and/or inadequate treatment prior to release).

In public health, framing excrement as an environmental problem has resulted in higher standards for contaminants in food and water, and more effective methods for treating water and food through heat, chemicals, and irradiation. On the farm, this has led to manuals that set out best management practices for a variety of farm-related activities including "Nutrient Management Planning,"

"Manure Management," and "Sewage Biosolids." Such best practices, associated with higher food and water standards, sometimes work,[1] but also, because of their costs, exert heavy financial pressure on already vulnerable smaller farmers. Even when these practices successfully minimize contamination of streams and soils, they are all, unfortunately, only of local utility, based on an industrial, linear view of causation in a non-linear, complex, natural world.

But what happens if we look at excrement as an ecological issue, rather than an environmental one? What if we focus on social-ecological relationships — on webs of life interacting with non-living environments across a wide range of spatial and temporal scales — rather than on inputs and outputs to some simplistic notion of an inert, localized environment? Ecologically, the impacts of manure are not always linear, and are poorly predictable. We already know, for instance, based on pandemics of infectious foodborne diseases, that excrement is being globally redistributed. But what (apart from public health concerns) does that imply?

Excrement is a package of partly digested foods with added bacteria and bodily fluids. The globally visible increases in shit represent a transformation of plant and animals that we use as food into a homogenous mash of fats, proteins, and carbohydrates that we rarely find palatable. If large mammals like people are the losers in this transformation of diverse megafauna and megaflora into shit, then, at least in the short run, insects and microorganisms are the beneficiaries. Does this mean we are moving from the anthropocene to the scarabocene? More

1 They apparently did not work to prevent the tragic outbreak in Walkerton in 2000. The farmer whose animals were the alleged source of the bacteria had implemented an Environmental Farm Plan.

generally, how do we know what happens to shit once it is separated from an animal's body?

The chemical components of excrement are not easily traceable on their journeys through the biosphere, and back through to their origins. Did this nitrogen come from soybeans capturing nitrogen from the air? Did this phosphorus come from erosion of rocks? Or did both come from animal feces? The piles of shit outside feedlots and intensive farms are the tips of the manure-bergs below ground, but, once the manure breaks down into its constituents in the decomposer webs, how do we actually know where it goes, or how much there is?

We can calculate manure volume based on how much an "average" person or cow or dog produces in a day, then count the people and cows and dogs, and estimate total amounts. We can weigh the materials that get flushed down toilets and piled up outside barns and see how that matches up with our "average" calculations. That will give us an estimate of how much is produced. But having done that, we are left with a lot of "dark matter," feces that does not show up in doggie bags or feedlot lagoons, but which we know, based on the number of animals in the world, must exist. Where does all that "invisible" shit go?

The landscapes and ecosystems we see are creations of the co-evolved relationships among organisms and the places they live. The decomposers, all those microbial lives we never see, are critical in this creative process. They take nutrients from plant litter, dead bodies, and feces and make them available again to new generations and to other forms of life. Even the most diverse systems experience collapse, reorganization, and renewal at small spatial scales. Forest ecologists Borman and Likens called such a landscape — with various parts of it at different stages of ecological development — a shifting steady-state mosaic. Besides studying dung beetles (a laudable activity), how can we begin to visualize, and perhaps better understand, the processes within which excrement is transformed and travels?

We can, in the first instance, consider the distribution of human and animal diseases that are attributable to fecal contamination of food and water. These diseases are important on their own terms. Multi-country epidemics of salmonellosis or diseases related to *E. coli*, however, are also important for what they tell us about where all the shit has gone in our social-ecological systems.

Beyond diseases, we can examine the life cycles of many shit-related and shit-eating organisms that are visible to the naked eye. From these we can see where the excrement is, how it reformulates into new life, and where its nutrient content is cycling. If we put together the patterns of disease, the networks that characterize the industrialized global agri-food system, and our observations of these macro-parasites, we are presented with some startling insights.

For example, the Surinam cockroach (*Pycnoscelus surinamensis*) is a burrowing insect in the tropics that is attracted to bird shit, not because it eats manure but because it eats plants. Because the birds' slurry of urine and feces is high in both nitrogen and phosphorus, a good place to find plants is near a pile of bird manure. While digging around in the dirt and wrecking greenery, the insect eats pieces of bird droppings containing the eggs of a parasite, *Oxyspirura mansoni*. Once inside the cockroach, the baby parasites go through a few changes and eventually take on a cyst form. They wait until the cockroach is eaten by a bird, and there, in the warm body of the bird, the larvae hatch and migrate up the esophagus into the pharynx. In as little as five minutes, they sprint up the tear ducts into the eyes, where they grow up in the shelter of the nictitating membrane, the so-called third eyelid. The mature parasites breed and make babies, which are washed in the bird's tears down the hatch and out through the cloaca — the bird's joint exit hall for feces and urine — onto another dung heap. Thus is dung transformed, and parasites dispersed to new locations.

Taking an ecological view, every parasite is not just an organism. It is also a bundle of nutrients, information, and energy: each organism, through its eating of the excrement, is an embodiment of excrement. We are, all of us, what we eat. The parasite cycle gives us a rich picture of how the nutrient essence of excrement, if not its form, can move through the ecosystem from the intestines of birds and mammals to the soil, to plants and insects, and back again to mammals. The parasites and their hosts and predators are, in fact, the re-embodiment of the deconstructed excrement; they are made out of shit.[2] The life cycles of these parasites are life cycles of excrement. The implication is that killing off any species, however small or obnoxious, closes off certain pathways of nutrient recycling and hence will affect our lives, sooner or later.

But even this characterization of the distribution of parasites misses an important part of the problem. We tend to think of issues like manure runoff from animal feedlots or human sewage from cities as local environmental or public health hazards requiring straightforward technical solutions. Build better sewer treatment plants. Create runoff lagoons and settling ponds for manure.

2 In this light, calling someone a shit should be seen as both scientifically accurate and complimentary.

Such waste management systems are helpful in the short run. They are the emergency doctors that save the person hit by a car. But they do nothing to prevent future accidents. What is left out of this short-term techno-picture is that cattle in the feedlots and people in the big cities are part of a system that is unprecedented in the 4-billion-year history of the world. Although the amount of biomass in the world remains more or less unchanged, there are more people and more livestock on the globe than ever before. Not only that, but these people and animals are being fed nutrients drawn from elsewhere, far away from where they live. In Canada or the U.S., for instance, human food and animal feed come from South America, Asia, or Africa.

According to the Food and Agriculture Organization,[3] in 2007 the United States exported roughly 57 million metric tons of corn; Argentina, 15 million; and Brazil, 11 million. That same year, Japan imported 16.6 million tons; Korea, 8.6 million; Mexico, 7.9 million; and Spain, 6.6 million. What, you might ask, does any of this have to do with excrement? Consider this: if we say corn is made up of about 10% water and 10% protein, then the top three exporters alone extracted the nutrient equivalent of 8.3 million tons of protein from their soils and shipped another 8.3 million tons of water. The top importers took in nearly 4 million tons of protein and an equal amount of water. In other words, a great many tons of nutrients (carbon, nitrogen, oxygen, and hydrogen, which make up the proteins and carbohydrates and fats; and phosphorus, which is part of DNA molecules), water, and other chemicals (including toxic heavy metals

3 The export and import figures are from the FAO website. I could have used any year, but these are for heuristic purposes, so the exact numbers are less important than the orders of magnitude. See http://faostat.fao.org/site/342/default.aspx.

like cadmium) were taken out of some ecosystems, shipped around the world, and injected as animal and human excrement into other ecosystems.

The reason we should give a shit, then, about crop exports and imports, is that moving all this material around leads to depletion of nutrients where they are extracted, and "pollution" of nutrients where the shit is deposited. Countries that say they are not exporting or importing water are ignorant. We do it all the time. It is just that the water is taken out of the ground, packaged in crops, and *then* exported or imported. Then, when these foods and feeds move through the people and animals and the undigested bits get dumped on the soils or flushed into waterways, they lead to public health problems, over-fertilization of soils, and contamination of water. We end up with deserts in the jungle and dead lakes on the plains.

And it is foolish to think that the nutrients extracted from Brazilian or African soils can simply be replaced with fossil fuel-based fertilizers or that the ecosystems on all our agricultural lands are not being restructured and changed in some fundamental ways by these practices.

All this human and other animal excrement is not evenly distributed across Earth. People are moving into cities faster than ever before in history; more than half the planet is living in urban centers and the numbers are going up. As people move into larger towns and make more money, they want to eat more animal products. To meet these wants, and to make money, agribusinesses have created great big farms, meaning that both human and livestock excrement is being crowded into fewer and fewer places.

In the 1950s, there were some 3 million pig producers in the United States. By 1965, there were about a million farmers raising pigs in the U.S., with an average of about fifty pigs per farm. By 1990, there were fewer than 200,000 hog operations, with an average of almost 200 pigs per farm. According to a study by the United States Department of Agriculture, "The number of hog farms fell by more than 70 percent between 1992 and 2004, whereas the hog inventory remained stable. The average hog operation grew from 945 head in 1992 to 2,589 head in 1998 and to 4,646 head in 2004. The share of the hog inventory on operations with 2,000 or more head increased from less than 30 percent to nearly 80 percent. Operations with 5,000 or more head held more than 50 percent of the hog inventory in 2004." Within this overall restructuring, there were also trends toward more specialization within the pig-rearing industry.

We have seen similar changes in farm size, and transformations in the structure of the agricultural systems, all over the world. These changes are driven by, and justified by, the usual excuses — or reasons, depending on your ideology — such as economies of scale to keep prices down, and, conversely it seems, increased production to meet higher demands for more meat from urban consumers with more disposable income. Pork and chicken production skyrocketed in South and Southeast Asia in the 1980s and 1990s, reflecting the "tiger" economies of those regions.

If we look at maps of where pigs and chickens are raised today, we can quickly see that livestock are not randomly scattered around the world. The Food and Agriculture Organization of the United Nations has published maps that show global distribution of different livestock species. On maps showing pig distribution, much of China, parts of Europe, and the midwestern United States are

Fauvist splashes of manure-colored rusty-brown. Poultry follow similar patterns of distribution on these maps, but are somewhat more dispersed. Cattle densities are highest down the central plains of North America, parts of South America (Brazil, Argentina), India, and East Africa, where the Maasai have herded cattle for centuries and consider themselves custodians of the global "herd."

Those splashes and splotches and smudges on the maps can be thought of not just as farms and animals; they are also pools and piles of animal excrement. My point here is not to point to those particular places where manure management may be a critical problem, but to point out that the distribution is neither random nor even. The impacts of livestock shitting (both in terms of where the ingredients of that shit come from, and in terms of where it goes) are heterogeneous and, depending on both the source and the end points, may be magnified by influencing critical places and species in ecological systems. Rain scattered over a vast landscape has a very different effect than that same amount of water channeled onto a few houses. One may be rejuvenating and healthy; the other can lead to floods and destruction.

By moving into cities in record numbers, and making certain decisions with regard to the foods we want to eat (less expensive, high protein, non-fat), we are changing the amount and distribution of excrement in the world. More than that, we are removing biomass incorporated in some species and redirecting it to others.

We might be able to live with this much manure if it were at least integrated into, and rejuvenating, depleted soils. But the manure is piled up in too few places, and these places suffer from, rather than benefit from, the nutrients and bacteria in the feces.

Since most of the rapid, unprecedented increases in feces globally have been related to birds (chickens) and domesticated mammals, most research has focused on issues related to managing those wastes. But wildlife, particularly those species that are well-adapted to landscapes transformed by people, can also be a source of significant fecal injections into local ecosystems. In some cases, this has been put to positive use. For instance, waterfowl have historically been important for fertilizing rice paddies in Southeast Asia, as well as providing food (through their feces) for fish in farm ponds.

We also know, however, from studies of bacterial DNA found in waters along the shores of Lake Ontario, that much of the fecal contamination of beaches near Hamilton, Ontario, comes from birds, in particular sea gulls and Canada geese. Each goose puts out a kilogram of feces every day. Ironically, these birds are often fed by people along the shore, who then later complain that the water isn't fit for swimming. Such "tamed" waterfowl populations are a major problem throughout North America.

Fish feces, often high in phosphorus, nitrogen, and calcium carbonate, are important in the aquatic food web. Calcium carbonate reacts with carbon dioxide, and has been used to counteract the acidifying effects of acid rain in aquatic systems. It may well be important in counteracting the acidification of oceans associated with global warming. Sperm whales, for instance, defecate about 50 tons of iron into deep ocean zones every year. This stimulates the growth of organisms that use energy (primarily from the sun) and combine it with carbon from carbon dioxide to produce organic molecules. The organisms that do this, called primary producers, sink to the bottom when they die, and are a significant form of carbon export to the deep ocean. The total population of 12,000 sperm whales in the Southern Ocean remove an estimated net 200,000 metric tons of carbon from the atmosphere annually. So the depletion of marine animals for pan-fried protein and "research" is likely a contributor to the frying of the planet. Conversely, as is the case with the concentration of livestock on land and the redirection of nutrients away from their historic ecological pathways, fish feces can be locally problematic as environmental pollutants leaking from fish farms into the surrounding ocean water.

In the twenty-first century, excrement has become more than just the normal nutrients and energy. It also contains many of the drugs, and drug-resistant organisms, that we and our domestic animals ingest. These are of more than just local environmental concern. The long-term ecological effects of the antibiotics and other drugs that are shit out by people and other animals are uncertain, but what we do know is worrisome. For instance, ivermectin is one of the most widely used antiparasitic drugs in sheep, horses, pigs, and cattle. Used globally to control gastrointestinal and pulmonary nematodes, lice, ticks, mites, and warble flies, this broad-spectrum drug has come to be considered something of a miracle among veterinarians and farmers. The drug works because it is extremely toxic to invertebrates, but not just to the ones we don't like. The activity of invermectin in the manure of treated animals can suppress dung-eating insects for up to a month and hence promote the buildup of excrement and the loss of nutrients to the plants on which the animals depend. Thus we are creating more dung and destroying the natural processes that would help us recycle it.

Recent experimental studies by a team of Danish and Canadian scientists identified acute and chronic risks to invertebrates in aquatic ecosystems from ivermectin at very low levels. Since invertebrates are essential for the functioning of all ecosystems, and we often count on the inhabitants of water systems to help break down the organic waste we dump into those systems, we are clearly flirting with danger.

The passage of veterinary and human drugs through excrement into ecological food webs is only one aspect of the issue. More insidiously, our global feeding and manure management practices, particularly the application of so-called biosolids to crops, have resulted in the redistribution of heavy metals such as cadmium, with poorly studied impacts on microbial populations in soils.

Many people have the naïve notion that we can dump our shit down the toilets and someone can simply remove all toxins and bacteria. Some municipalities have made a lot of headway in doing this, but the processes are expensive and time-consuming, and can be overwhelmed by the sheer volume of shit we produce. Whereas night soil is untreated human feces applied to land as a fertilizer, sewage sludge is a combination of wastewater and the solid materials left after wastewater is treated. Wastewater is everything that gets flushed down the sewer; it is mostly feces, but can include anything else that people or industries decide will fit through the pipes. Cities treat that shit, trying to keep "useful" elements such as nitrogen, phosphorus, copper, iron, molybdenum, and zinc at reasonable levels, while removing toxic chemicals and heavy metals. Some of the treatments may involve applying bacteria that can extract these substances. The materials that come out at the end of this process are called biosolids.

These biosolids — providing they meet certain regulatory standards, particularly for heavy metals — are then applied to agricultural lands. Depending on where you are in the world, anywhere from 30 to 50% of biosolids are applied to agricultural land; the rest is incinerated or goes to landfills. Putting biosolids on agricultural lands is seen to be a better (more ecologically sustainable) option than incinerating the material or dumping it into landfills.

Whether this practice is dangerous or not may be the wrong question to ask. There are always risks and benefits; they just vary in terms of who risks, who benefits, and over what time scales. The return of nutrients to agricultural soils would seem to be a good thing, in keeping with what we understand of ecological cycles. However, since we know that certain root crops, such as carrots, take up cadmium from the soil, we need to look at the foods coming from the soil. A range of variables, such as the source of the biosolids (industrial, agricultural, human waste from urban populations), the nature of the soil, the local climate, and the type of plant all play a role in whether those metals are taken up.

So, not all recycling is good. The manure in the compost heaps on farms, or in the manure lagoons, or the fecal waste that ends up in city sewers reflects the origins of the food and feed that went into the mouth and came out the butt. When we ship soybeans from Brazil to Sweden, feed them to animals, and collect the feces, we are harvesting both water and nutrients from another part of the world and transferring them to a new ecosystem. In her doctoral work with Carl Folke of the Stockholm Resilience Centre, Lisa Deutsch reported that more than 70% of manufactured animal feeds for pigs, chickens, and cattle in Sweden depend on imported ingredients from places as far-flung as Southeast Asia and South America. These kinds of translocations of feeds (read: nutrients, water, energy) change ecosystems as seriously as when we shipped cane toads to Australia from the Americas (to solve pest problems), and rabbits from Europe (for hunting). The catastrophic impacts of both these translocations are justly infamous and well known. In the case of feeds, the impacts appear to be less dramatic, but they may be even more important. We are changing both ecosystems — those where the feeds originate and those to which they are shipped. Putting the shit on the land in Sweden may look like recycling, but it is actually a massive translocation from Brazil of water and nutrients, including heavy metals and other unwanted "soil amendments."

Similarly, when cattle in Canada are moved from small cow-calf operations in Ontario and the prairies into Alberta feedlots where they are fed corn from the midwestern U.S., we are shifting energy and nutrients back and forth from woodlands to prairies and back again on an enormous scale. When water buffalo in Nepal are moved to be slaughtered for meat, we have removed nutrients from the countryside, where they are needed, and dumped them in the city, where they are a problem. We are physically restructuring global ecosystems by the way we feed animals, by what we ourselves eat, and by what all of us shit out.

Every life has a cost. Every hectare of land cleared to raise or feed pigs or chickens or cattle or people is a hectare of land unavailable to feed other species. The transformation of diverse landscapes into excrement is an insult added to this injury inflicted on the planet. Given the human population on this planet, even a vegan diet is associated with the deaths or slow disappearance of millions of animals through habitat loss. We might get away with some of this if we anticipated that the Earth is a stable ecosystem and that the future will be like the past. This is unlikely. Since the future is uncertain, we need as many species, ecosystems, ideas, and cultures as possible, to keep our options open.

In the end, nature doesn't give a shit what happens. We care, however, about the state of the world we live in, and whether it will be a congenial place for the next seven generations.

Global trade in human food and animal feeds — more generally, our particular ways of manipulating the services ecosystems provide — represents an unprecedented transformation and re-distribution of organic matter in the biosphere. We are taking a brilliantly complex diversity of animal, plant, and bacterial species and transforming them into a disordered mess of bacteria and nutrients. We are transforming a wonderful, complex planet into piles of shit.

CHAPTER 8

MAKING SENSE OF EXCREMENT'S WICKED COMPLEXITY[1]

The list of excrement-related problems is long and growing by the day: fecal-transmitted parasites in Nepal; excrement-associated foodborne disease outbreaks in Europe and North America; epilepsy and other neurological disorders, some of which are fecally transmitted parasite infections (recall the *Toxoplasma* and *Taenia solium* stories); nitrate pollution of water in the Netherlands; bacterially contaminated water in cities the world over. Also related (less directly, but still in demonstrable, substantial ways), food insecurity and famine in the horn of Africa and rapid, dysfunctional urbanization in most parts of the world are

1 Details of the general theory and practice set out in this chapter and the next are addressed in much greater detail in *The Ecosystem Approach: Complexity, Uncertainty, and Managing for Sustainability* (New York: Columbia University Press, 2008); *Ecosystem Sustainability and Health: A Practical Approach* (Cambridge: Cambridge University Press, 2004), and *Ecohealth: A Primer* (Victoria: Veterinarians without Borders/Vétérinaires sans Frontières – Canada, 2011. Available as a free download at https://www.vetswithoutborders.ca/get-involved/resources).

providing a justification for large corporations to centralize and globally trade both food and excrement.

The world seems to be tumbling out of control, and the old scientific, technical, and political agendas seem powerless — and lack the economic, moral, and intellectual resources — to respond effectively. We simply cannot set a priority of diseases and problems and, one by one, go down the list and get rid of them. The list is endless. Worst of all, the kind of solutions we devise to some problems — say, of food shortages — create as many problems, such as creating ideal conditions for emergence and spread of new diseases, as they solve. Treating animals for parasites may destroy the landscapes that support those animals.

At the core of the wicked mess of shit, food, and ecological sustainability is a challenge of theory. We have developed ad hoc solutions, using a Henry Ford, linear, industrial model of nature. This theory works in a factory, or in a laboratory, but wreaks havoc in the world outside those confines.

"Normal"[2] scientific (clockwork, industrial, linearly causal) assumptions about the world provide no formal way of integrating all the relevant pieces, or even to even ask the important questions. It is not an accident that shit and science have the same linguistic roots. How might particular solutions to food security affect ecological sustainability?

2 There is no good word to describe the kind of limited scientific inquiry I am talking about. Reductionist, linear, and industrial have all been used. This kind of science is often associated with laboratory experiments where "all other things" are assumed equal, or under control. In this view, science is comprised of multiple disciplines, each with different rules of evidence, and the "whole" is, in some simplistic way, considered to be the sum of the parts. I suppose I am talking about what Thomas Kuhn called "normal science," and I shall use that as a kind of shorthand. I am not dismissive of that kind of science. It is excellent for answering some kinds of questions (measuring drug efficacy in hospitals, assessing chemical hazards) but is weak at addressing the questions I am trying to get at in this book.

How might certain solutions to manure management problems affect climate change or food security? How might some responses to veterinary or medical problems actually undermine long-term health? How do we deal with the grief that necessarily comes with change, no matter how important or "good" the change may be? Can science be "redeemed"? The answer to this last question is: yes, if we take a broader view of science as a way of generating knowledge grounded in the "real world."

If we cannot set a list and solve the problems, one by one, what are we left with? In conventional social development and normal scientific terms, not much. Researchers do excellent, independent disciplinary work, based on what the philosopher of science Thomas Kuhn called normal science: a science where every discipline has its own rigid rules of practice, acceptable evidence, and quality control. The scholars then hand the result over to politicians, whom they expect to be much more thoughtful, insightful, integrative, and visionary than any university scholar. We then lament that the politicians make decisions using criteria based on someone else's evidence rather than our own.

The policy makers, even the most thoughtful among them, even those for whom "One Health" (integrating the health of people, other animals, and ecosystems) is at the top of their agenda, are faced with the challenge of trying to integrate the social determinants of health, the environmental determinants of health, and their limited budgets. Manure management is competing with child poverty and maternal health and population control and disappearing whales and pesticide pollution and avian influenza and cholera and hunger and salmonellosis and childhood obesity and car accidents. The list, indeed, is endless, resulting in one plea after another for more money for more programs.

Although fraught with danger and uncertainty, the task is not hopeless, because, fundamentally, the individual problems are part of one big, wicked problem. If we can get our heads around that, we can start to come up with viable solutions. We are a pretty ingenious species.

Let us go back to some basics of science and knowledge. Scientists are hunters and gatherers. Instead of fruits and nuts, we track the scats of reality in a confusing, deep forest we call the world. We do not actually *know* the world; we perceive it with our senses — sight, smell, touch, hearing, taste. These inputs create a set of neural connections that function as an internal mental model of "what is out there." This mental model determines our behavior, and our survival.

Sometimes individuals are wrong. Sometimes a person is psychotic, or their neural connections or biochemical messengers get mixed up, or they take a singular case to be of general applicability, or their culture biases them to see some things and ignore others. Sometimes an animal ends up in the wrong place at the wrong time because the landscape on which its brain connections were based has been changed. Mistakes can be costly. A prey animal could be eaten, or run over by a car, or fly into a wind turbine, or a person could eat food or drink water contaminated by feces that contain cholera or *E. coli* organisms, or the markets for which swine farmers have been raising pigs might suddenly disappear because of public panic over a new strain of influenza.

We overcome these deficits by sharing information, by critiquing each other based on our perceptions, by telling and hearing our different narratives that we use to make sense of the world. "Based on my experience," we might begin to say, or "based on this study we did." This is no more, nor less, than a hunting story told around a campfire, a way of constructing meaning from facts.

Reality scat-gathering is only a small part of science. Building meaning around the facts, in a collective, open manner, is what differentiates a scientist from a junk collector. In this complex and uncertain world, good science is a systematization of the process of storytelling about perceived facts. Good science, the best science, is a way to share experience, to offer alternative explanations, to project future possibilities based on past experiences.

What does this mean in practice? The universe in which we live is characterized by a very rich set of connections, and what we see around us is open to a wide variety of influences from its surroundings. Some mathematically inclined complex systems theorists believe that complexity can be explained and "re-created" virtually using a few simple calculations that repeat themselves. However, most of us who work with complexity as it relates to addressing public health and environmental problems would challenge that assumption. Quantitative modeling can provide some interesting insights into the spread of a disease or the movements of nitrogen in the soil. However, modeling the complexity of the world — particularly if we include that most troublesome species, ourselves — in any comprehensive and definitive way is impossible. John L. Casti, who has written extensively about complexity research, calls it the "science of surprise." If one takes this position, then only by accommodating many different perspectives can we begin to gain insight into some of the surprising ways in "how the world works."

In practice, the complexity we see in the world is a function of the nature of the world itself, us who observe the world, and the questions we ask. If we ask how to fix a broken watch, or how to collect a stool sample from a dog, we can think of the watch or a dog in fairly simple mechanical terms, and we do not need to invoke notions of complexity. If we ask about the function of watches in society, or the social, political, economic, and ecological relationships required to acquire the resources, materials, and skills necessary to build a watch, we need to invoke complexity. Similarly, if we wish to save people dying of cholera, we have the relatively straightforward, albeit challenging, task of providing them with potable sources of fluid replacement. If we wish to prevent cholera epidemics, we are faced with complex, interacting political, social, economic, biomedical, and ecological forces.

How do we wiggle some space here to make important, often urgent, decisions? For practical purposes, we can take some key sets of relationships, look for basic principles or tendencies, and make at least reasonable projections as to outcomes, given certain inputs. Once we have tried something, we ask, did we see what we expected? In doing this, we are both engaging in systems thinking and simplifying the world in important ways. We should never confuse our models — however complex they may be — with the complexity of the real world. The only way to begin to approximate an understanding of the real world is to bring together as many perspectives as possible and try to gain a collective picture. Again and again.

If we think systemically about the world, it seems reasonable to propose that everything is connected, and that therefore everything we do has unintended consequences. However, not all things are connected or related equally: relationships have different strengths, and sometimes the links are spread widely across space and time. For instance, the choices European and North American consumers and farmers made with regard to increasing food supplies after World War II have had a vast array of unintended social and ecological consequences, many related to creation and disposal of manure. These consequences were only apparent in hindsight. Few would have predicted today's world based on what people observed fifty years ago.

How then do we parse our thinking about the world systemically so that it becomes useful to us? First of all, we can generally classify our thoughts about systems into three groups: simple, complicated, and complex.

If I arrive in an emergency room after a car accident, I want to be treated by experts who know what kinds of fluid replacements to administer, how to insert a needle into a vein, how to fix a broken leg, and so on. If there is a spill from a livestock manure-holding lagoon, I want to engage good engineers who understand how to fix the spill as soon as possible. Drawing boundaries around these problems and thinking about them in fairly simple terms is the quickest way to solve them. I don't want the physician in the emergency room to ask me about my life's goals.

From a scholarly point of view, we can improve our understanding of simple systems by gathering more knowledge and developing good, relatively simple models. Practically, we seek good education and training, we want a hierarchy of command, and efficiency is good. We need to remember that this "simple system" version of a body's condition or a manure spill is a function of the questions being asked and the problems we want to solve. In the face of an emergency, or even when a computer or a watch needs fixing, asking about the meaning of life or the role of technology in modern life is not helpful. Simple system views are starting places for asking questions. For a publically engaged scientist, they are *never* places to end.

Some issues evoke a more complicated systems view than fixing a watch or a bone. For instance, I can create mathematical quantitative models that predict that, if I do certain things in a certain way, I can land a craft on Mars, or dispose safely of certain amounts of manure on a piece of land with known soil type and in an area with relatively stable rainfall. Actually, I can't really do that. More to the point, I can find experts who can. Working with complicated systems is highly expert-reliant. Because the mathematics and modeling are difficult — because there are so many relationships we are trying to keep track of — we need redundancy. We need to ensure that there are checks, balances, and options for Plans B and C.

Good education and training are important for tackling complicated problems. However, unlike for simple systems, effectiveness (achieving goals) is more important than efficiency. It was more important to actually land the *Curiosity* on Mars than to get it there as fast as possible using the least amount of fuel (although that was obviously one of many issues involved).

If we assume that the world is relatively stable, then we can imagine a great many problems to be solvable using complicated, multi-expert thinking. The planets and their gravitational forces were not likely to shift radically during the voyage of the *Curiosity*. In a reasonably stable climatic and political climate, biological diversity in farming or a mix of health-care delivery personnel in a medical system offer buffering. Redundancy gives the system some resilience in the face of change and stress.

Unfortunately, our best systems of checks and balances break down if the world becomes unstable. In the face of rapid climate change and political and economic instability, we are hard pressed to predict which crops will grow well, what the requirements for health care delivery might be, and which manure management systems are best. Indeed, in a world of incredibly diverse ecosystems and cultures, that complicated systemic thinking provides poor guidance even if the world is relatively stable.

As the context we live in becomes unstable, and as we move away from the illusion of a one-size-fits-all world, the problems we see as complicated are rapidly transformed into complex challenges. This is particularly true in city planning, agricultural and food systems, and the manure problems associated with them. Economies of scale (for both food production and manure disposal) make efficient use of certain kinds of resources under stable economic and environmental conditions, but are very brittle in the face of change. If everyone is growing corn, and the price of fossil fuels goes up, or the market for corn collapses, the system cannot adapt. People might go hungry because all the corn is used for fuel or, conversely, because they can't sell their crops.

Large slaughterhouses that require high through-put of animals have had to shut down completely in the face of border closings or sudden changes in market access (due to BSE, or foot and mouth disease, or fluctuations in the price of oil). This can have cascading, devastating effects as farmers cannot even service local markets.

Suddenly we have shifted into a disorienting world where many of our predictive models are not helpful. We are like polar bears lost in a tropical desert. We are like fastidious office workers up to our necks in shit. This is when we need to think about complexity.

Complex systems are descriptions of complexity — attempts to describe the world as we live in it and experience it. Precisely because we are trying to understand incredibly tangled and unstable relationships, there are many such descriptions, or models, possible; different observers will see different things in the world and model them differently. Although mathematical models are useful to explain certain events (such as pandemics, climate variability, or the ecological and climatic impacts of manure creation and disposal), there is no single mathematical model that can pull all of these together with human behavior and allow us to predict what will unfold in the future. Raising a child, for instance, is not like sending a landing craft to Mars: it is much more complex and difficult. Managing a sustainable food and shit system or land use in a watershed where industry, human settlement, wildlife, and food production are vying for space is similarly complex.

Complexity theory has many facets, and it is not my intention to explore all of them here. I am particularly interested in those that might help us address the wicked problem of excrement. These include some basic elements of systems thinking (that is, looking at relationships and interaction), scale (both temporal and spatial), and multiple perspectives.

If, just for argument's sake, we separate the world we inhabit into social systems and "natural" or ecological systems, and we assume that these exist as separate entities, we can imagine how human activities such as eating and manure disposal change the context in which we live, and how those ecological changes in turn change human societies. We change human social systems by the way we design cities, produce and distribute food, the way we handle manure, trade, travel, create transportation systems, build dams, and harvest trees. These changes in social systems in turn change the flows of energy, materials, and information (genetic, behavioral, cultural) through ecological systems. These changes in flows then create different outputs (changes in water flows, emerging diseases because of encroachment on wild areas, generating heat islands from paving areas or cutting down trees). Circling back around, these different outputs put pressure on social systems to adapt. Suddenly we are faced with pandemics of new diseases (SARS, for instance) or storm surges in major cities (such as tropical storm Sandy), for which our infrastructure is not prepared.

Related to this characterization of how social and ecological systems interact is the notion of emergence: when many variables interact, something entirely new may emerge. Or at least, if we look at the world in different ways, we may see something new. Think back to the exercise of looking through the microscope and the standing back that I introduced in the chapter on evolution. Single cells interact to create multi-cellular animals in an ecological context, but we don't see the animal until we step away from the microscope. We don't see a forest or a grassland until we step away from the tree or the blade of grass. Sometimes some genuinely new thing comes into being. Who would have predicted a boreal forest based on the first unicellular life forms? Who would even predict the flows and boundaries of such a forest based on looking at a couple of trees?[3]

The late James Kay, a world-renowned systems design engineer, ecologist, and physicist, explicitly teased apart how interacting changes in social systems and ecological systems were linked through flows of energy, materials, and information. By studying a wide range of situations, from urban parks and rural agriculture to industries and large-scale protected areas, and combining what we can visibly see with a deep understanding of the complex energy flows

3 A common example given for this is the birth of the internet, which was not predictable based simply on the known pieces of communications technology and human behavior.

in ecosystems, he was able to extract some general principles. Social systems, when faced with an energy gradient across their boundaries (inputs of energy by fossil fuel, or embodied in people and animals) build material structures (armies, businesses, farms, skyscrapers) to use and dissipate that energy. The structures, built from the materials available and at hand, physically change the landscape in terms of energy, materials, and information (genetics, biological structures, learned behaviors, acquired knowledge). This changes the natural systems that formed the context and inputs for the social systems in the first place.

Suppose people wish to decrease the price of chicken and make it more available to people with low incomes; these are social goals, the result of changes in political and business organizations. Using massive inputs of fossil fuels, people accomplish this by completely restructuring the physical landscape of agriculture to accommodate the growing of grain crops to feed the chickens, as well as to provide consistent, climate-controlled housing for the birds. This changes the way rural communities are organized and how they relate to cities, and creates opportunities for bacteria to find new pathways into social systems. So, as long as the sun shall shine or we mine stored solar energy (fossil fuels), social-ecological systems are faced with energy gradients and are caught in endless feedbacks of re-structuring.

In the same way that a human body is something more than the sum of its chemical interactions, an urban society with certain resource demands creates something more than its constituent parts. The accumulations of manure, and the new diseases that we see are part of that new entity, and may not be controllable without changing some fundamental organizational structures in agriculture and land use. Seven billion people on Earth create a qualitatively different social and ecological dynamic than one billion.

We have talked about how human communities interact with, and change, the ecosystems in which they are embedded. This does not only happen locally. These changes happen at every scale (individuals, households, neighborhoods) and across scales (individual car use collectively changes the global atmosphere).

The kinds of resilience, integrity, and health that interest us when we pursue research into sustainable development can only be understood in terms of interactions across many geographic and organizational scales.

Depending on which research literature you read, such systems are variously called: complex adaptive systems; self-organizing, holarchic,[4] open systems; or social-ecological systems. In general, the "complex adaptive" and "self-organizing" refer to the fact that healthy or resilient natural systems will, in the face of various external pressures, maintain their essential functions. Species may disappear and temperatures may rise, but as long as the remaining organisms can still organize themselves to build structures to deal with increased incoming energy, and can recycle essential elements such as phosphorus and nitrogen, they will survive.

What does that re-structuring and self-organizing look like over time?

It used to be said — based on limited observation of ecological changes — that all ecosystems move through succession from immature to mature states and somehow stay there. Now the situation does not appear quite so

4 James Kay called these SOHOs. I'll talk more about holarchy and holarchic later.

simple. The Resilience Alliance, a worldwide network of researchers investigating issues of linked social-ecological change, has studied sustainability and change in ecosystems at multiple scales. Their original conclusion, building on the groundbreaking work of Canadian ecologist C.S. "Buzz" Holling, was that every ecosystem went through four phases of development.

In phases 1 and 2, various species compete and cooperate to create structures in particular ecosystems. These phases are called exploitation and conservation; at some point, perhaps when the system is "over-connected" it becomes brittle, collapses, and goes through a process of reorganization (phases 3 and 4). Holling first called breakdown and renewal in ecosystems "creative destruction" but, apparently under attack from scientists for whom the creative and intelligent use of language is a foreign concept, later renamed the turning points to "release" and "reorganization."

On a small scale, the creative destruction in this cycle was reflected in small forest fires, small patches of spruce budworm destruction, small outbreaks of disease. After the small fire or outbreak, plenty of genetic and nutritional material was available nearby for renewal.

When the local fires or outbreaks were suppressed, leaving lots of dead wood and disease-susceptible trees around, we scaled up the problem. The result was huge fires and disease outbreaks. The point is that if we remove some species from a natural system, or change the relationships in that system from how they have developed over time, we need to carry out the functions of the species we have removed. We need to relate differently to the ecosystem we have altered, to deliberately perform the ecological functions performed by the fires and outbreaks (removal and recycling of old, dead, or brittle materials, plants, and animals, re-seeding, fertilizing). Some functions are more difficult to replace than others. If we inadvertently destroy dung beetles, who will perform their functions? If we remove seed-dispersing bats or birds or pollinating insects from the equation, what happens to the options for renewal?

With great success, farmers have harnessed this process by planting new seeds, killing all the plants and animals they don't like, harvesting the "mature" crop, destroying the remaining plant structure of the landscape (plowing), and, finally, renewing the fields with seeds of the plants (perhaps new, genetically modified ones) they want. These farming activities are based on anticipation that next year will be like last year. The farmers' anticipation is, however, different from, say, a marine fish, whose life cycle assumes certain stable temperatures and food availability. The farmers change the present (e.g., plant more corn) in anticipation of both a stable climate and a higher price for corn. If all the farmers plant corn, they change the landscape, contribute to an unstable climate, and drive down the price of corn. I'll come back to this later, because there is an upside to this as well.

Local events interact with larger-scale events across spatial and temporal scales, changing how life evolves on the planet. Large-scale changes — in climate, agricultural practices, or urban land development — alter the amount of exergy,[5] and the kinds of information and materials

5 Exergy is a term some systems engineers use for "useful energy." Energy is not created or destroyed, but some forms of energy (meat) seem to be more useful to us than others (shit). Exergy thus considers both the inherent energy (what a physicist might measure) and the context (how we wish to use that energy). Actually, shit still has a lot of energy, and if we change how we use it (the context), we can use that energy, instead of throwing it away.

(nutrients, elements, seeds, and animals) available for local renewal. We are then caught in a whole set of complex feedbacks. Genetic "innovations" and "revolutions" and "counter-revolutions" at a local level can radically change what happens at larger scales. These local changes might be related to, for instance, antibiotic use and the evolution and spread of resistant bacteria, or changes in livestock-rearing practices contributing to climate change and the global cycling of nitrogen, phosphorus, or water. If by changing some of the components we actually alter the relationships that provide the context within which species live and evolve, then species start to disappear. They become like refugees from "the former Yugoslavia": their country disappears. The species literally have no job to go back to.

It used to be suggested that you could rebuild an ecosystem if you saved all the pieces. We're no longer so sure that this can happen. Gene banks are interesting, but they'll never get us "back" to where we were before, since what genes "do" is determined by the context in which they occur.

What is important to note for our consideration of excrement is that the reorganization and renewal of any social-ecological system, in the face of ongoing, incoming energy, depends on the materials and information that are available to it. When we see large piles of manure outside a feedlot or chicken barn, or large manure lagoons outside a pig farm, what we are seeing is a loss of information. We can only regain that information, and achieve some degree of resilience, if the manure is an integral part of the relationships among organisms in an ecologically diverse landscape.

Large-scale industrial agricultural enterprises and the big cities they serve may be efficient at processing energy, but they are brittle in the face of major change. They do not have the diverse internal resources or connections to use the information in adaptive, self-organizing ways. The information they might have used is all piled up in a dung heap.

There is no single way to think about all of these interacting changes, but there are a variety of useful possibilities.

The Resilience Alliance researchers have termed the multiple, multi-scalar growth, collapse, and change that we see around us "panarchy." Sustainable development is seen, in panarchy terms, as a process of innovation and memory, revolt and stability. The innovation and revolt tend to occur locally, from the ground up. The memory and stability are encoded in long-term climatic and cultural patterns.

Philosopher Arthur Koestler talked about nature as being like the two-faced Roman god Janus. Each organism, person, family, or watershed is both a whole, with its own internal rules and feedbacks, and a part of something larger. Koestler called each of these whole-parts a holon, and called the overall structure a holarchy. In recognizing both the structural insights from Koestler and the dynamics of panarchy, Henry Regier, environmental scientist and former director of the Institute of Environmental Studies at the University of Toronto, introduced the term "holonocracy." Holonocracy embodies a way of interpreting nested social and ecological changes and implies a new way to think about management and governance based on those observations. Democracy is generally interpreted as being "flat" — populations of people with equal standing before the law. In an autocracy, one person has unlimited power over everyone else. Often, the debate in politics is based on a binary view of the world: the individual versus the state.

Viewing politics and governance only as they relate to individuals and states is not helpful in solving the challenges of living in an overcrowded, unstable, interactive, extremely puzzling world. What I find more helpful is a view that sees micro-organisms as part of tiny micro-landscapes (in an intestine, for instance), which are part of larger communities (animals and plants on a landscape) and ecosystems (cycling nutrients), all the way to the biosphere and the universe. Similarly, the individuals within which the bacteria live can be seen as members of households, who are parts of communities that work together in larger geographic regions, all the way up to the globe. Each unit (bacteria, animal, turd) interacts with, and is indeed part of, larger units of nature, and is comprised of smaller ones, all the way down to the invisible bits that physicists look for.

Regier's holonocracy, together with panarchy and Kay's models, give us frameworks for how to think about and manage all this shit, not just to describe it. Tailoring options to fit local conditions creates problems for policy-makers and managers, who want a "one-size-fits-all" solution. In natural systems, every solution to excrement recycling is very finely tuned to local conditions. Well-meaning environmental laws designed to control manure use in industrialized systems are a double-edged sword. They might work for big feedlots, but trying to force small farmers and local communities living under diverse conditions to adopt solutions and regulations suitable for large-scale industrial operations represents a kind of top-down dictatorship. This can put many smaller operators out of business and discourage neighborhood-based actions in cities. The expense of some systems only makes sense at large economies of scale; furthermore, this approach can suppress innovation to find more effective, less expensive, locally adapted solutions for dealing with manure. If we think in terms of a holonocracy, the purpose of national and global policy is to provide rules, support, capacity, information, money — to provide whatever is needed to help local communities and ecosystems to thrive — and to supply resources for renewal when local collapses occur.

How can we begin to assess the resilience of social-ecological systems? How can we "see" the movement of materials, information, and exergy? One way is to use, heuristically or metaphorically, observations on the relationships among plants, animals, and other organisms. Ecosystems are made up of (1) primary producers, which use sun and physical factors in the environment to make food; (2) consumers: animals (including ourselves), and bacteria that use, combine, and transform producers, food made by producers, and/or other consumers; and (3) decomposers: micro-organisms that break down dead organisms and organic waste. These three types of organisms are the physical manifestations of the dynamics of ecosystems; they perform the functions of cycling nitrogen, say, or water, even as they perform the functions of building structure and systemic resilience.

The life histories of parasites and insects and plants reveal a set of pathways by which the survival, growth, and adaptation of bacteria, insects, plants, and animals are linked across time and space. They reveal that, indeed, we are all that invisible shit made manifest, and that the natural order of life is based on shit.

In evolutionary terms, ecological systems "self-organized" and stayed resilient by reusing, repackaging, and recycling nutrients in forms that we would call waste, or shit, if we are talking specifically about animals. If we follow through on the general argument that more resilient systems are better at dissipating energy, then the more pathways for energy use a system can develop, the more it can keep on creatively (and thoughtlessly) organizing itself, and the more resilient it is. Can we mimic this diversity of pathways, and reclaim some measure of resilience, as we invent new kinds of living spaces and ways to feed ourselves?

I believe we can. If we understand the world, and our place in it, based on a more complex view of reality than we have been taught, if we go beyond normal science and politics to accommodate multiple perspectives, and multiple and complex systemic models, we can co-create a sort of global, open-ended narrative. Unlike the farmer who depends only on corn, or the city whose health is at the mercy of one or two treatment plants, a viable flush system, and a regional power grid, we want to create a situation where we have the capacity to anticipate a variety of futures.

If the world is a place for which we have no single description of a problem, and where the multiple perspectives held by microbiologists and anthropologists, shamans and epidemiologists, ecologists and economists, farmers and sex workers and musicians each have some validity, some truth, then how can we proceed? How do we accommodate multiple, jostling perspectives and maintain quality of information, and not just accept any flakey idea that comes along? How do we get all these diverse and often antagonistic people engaged in telling a collective, adaptive, diverse story?

Engagement by all these people in the process of generating knowledge and creating a working understanding of the world is not simply a way to facilitate political agendas, or promote new products, or gather people's opinions. Citizen engagement and multiple perspectives are an essential part of any new way forward, out of the dung pile and into the fresh air. Everyone brings to the table different kinds of evidence in the same way that a personal narrative history, a clinical examination, an epidemiological study of varying populations at risk, and a laboratory test inform a clinical medical decision. Participatory-action research and its relatives are not merely ways to get people involved and to develop more effective solutions, as important as that is.

Public engagement is essential to do good science in a context where the facts are in dispute, values are contested, knowledge is uncertain, decisions are urgent, and decision stakes are high. Philosophers of science Silvio Funtowicz and Jerry Ravetz have called this post-normal science, because it is not about overthrowing or replacing the many paradigms of science, but of accommodating them. It is a democratic kind of science, open to different *kinds* of knowledge and differing interpretations of the facts.

Anything less than this leaves open too wide a possibility that we are all psychotic, believing that we can manage to fit the world into any one single vision. That we believe in final solutions to disease, or poverty, or economic disparity. Anything less than this new science leads to the tyranny of public health warring with the tyranny of free trade economics warring with the tyranny of environmental policy.

But what is the point of this new science? Democratization of knowledge? Yes. Better science? Yes. But to what end? For the obsessively curious, this may be enough. But those of us who maintain illusions of changing the world, who want to move from knowledge generation to policy and action, we need a goal, something to inspire us and give us focus.

Resilience has been offered up as a goal by many ecologists. I have no quarrel with that. Another useful way to articulate the goal of this new science — one that can accommodate, in particular, the challenge of excrement and get the attention of people who otherwise don't care about "the environment" — is health. Not just any health: One Health.

Rene Dubos, one of the great public scientists of the twentieth century, described health as a mirage, since it was forever shimmering somewhere on the horizon but never attained. But health is attained every day and reinvented and rejuvenated in every culture. It is not a mirage; sensitive as it is to what people eat, who their friends are, the amount of sunshine they are exposed to, the weather in general, the quality of water, health is a renewable way of being. We all want to be in a state "of complete physical, mental, and social well-being and not merely the absence of disease or infirmity," which is how the World Health Organization (WHO) defined health in its 1948 constitution. Some of my colleagues have suggested that this sounds like an orgasm, something to enjoy periodically, but not a sustainable state.

Even with that caveat — some would suggest especially with that caveat — most of us would agree with the proposition that health is a good thing for all species, including people, and the planet we share. This is what the World Health Organization, the World Organization for Animal Health, the World Bank, the United Nations, and multiple government and nongovernmental agencies around the world call One Health.

One might well ask, when we speak of One Health:

Whose health? North Americans? Africans? Asians? Rich people? Poor people? Our children's children? Most importantly, individual health or population health or community health? Because it is possible to pour millions of dollars into saving and vaccinating babies, and, without equal attention to education, meaningful work, food security, we may make things worse — much, much worse — for those children when they grow up. Can we pull back from the catastrophic brink that an already overcrowded planet has pushed us to?

If we pour millions into keeping older people alive, what does this mean for the draw on resources needed for those children to having meaningful, fulfilled lives? Are we not then promoting global poverty, youth unemployment, and its natural, well-demonstrated consequence, war? If we put flush toilets and running water into slums so people can wash and drink, while drawing down already precarious water tables, what does the future hold for them? Or if we advise people in the Amazon to eat fruit-eating fish, rather than carnivorous fish, to avoid mercury poisoning, will we remove a source of dispersal for fruits trees, and what will be the consequences of that? If we focus only on food production and manure management as global goals, without paying attention to how we achieve them, will we not make things worse? Haven't we already done so?

And if we ask these questions, we have to deal with tragedy, loss, and suffering. We have to deal with all the cultural rituals, music, poetry, religion, with which we manage our grief. Because, in the long run, health for all means grief for many.

Yet health is a sufficiently robust idea that it somehow continues to engage us, and keeps us talking, even as we struggle with the details. And in talking together, we are already creating the social bonds that foster health. One Health provides a space for us to argue about who we are, who we want to become, and about all that shit out there that we are unsure how to "manage."

Health, in a global sense, expressed somewhat differently in each culture, remains an overarching goal, universally (if imperfectly) understood and desired. Yet why is it that this ideal of health seems to keep slipping away from us? Are we thrown back to the details of who and where? Of manure problems versus regional famine? Of babies against old people? Of climate instability versus meaningful work?

Conventional technical and scientific wisdom might have us believe that how we achieve goals is irrelevant. All that matters is a diagnosis and a technical treatment. And yet we know, by and large, that this doesn't work.

We can tell people not to smoke, to eat better, to exercise, to vaccinate their dogs, not to let their dogs or chickens run in the street, not to dump their manure where they please, and on and on, yet, in the full knowledge of what they are doing, they resist changing their behavior. This is not because they are stupid. This is because people are working with multiple demands on their time and attention, and the world does not function like a factory or a clock, with linear cause-effect chains. The world is a mess of interactions, and every ecosystem and social-ecological system we can imagine about the world is a simplification, a simplification from which we learn, but which nevertheless is incomplete.

The only defensible way to act is by engaging with others, sharing information, changing our minds, being humble in the face of an amazing universe.

How we act is as important as the goals we set. With One Health as a goal, complexity as a theoretical base, and post-normal science as a guiding principle for linking science with action, we can move forward with telling the story of our lives on this amazing and sometimes infuriatingly contrary planet. This way of framing and addressing wicked health issues, often referred to as ecohealth (short for ecosystem approaches to health), has led us from a manure pile next to the barn to linked social-ecological systems to a new way of doing science. Now is the time, I think, for us to think about solutions.

CHAPTER 9

KNOW SHIT: A WAY FORWARD

I think the hardest thing to feel is that the shit is purposeless. This is at the heart of the emotion we call "despair." Despair is an existential emotion. It occurs when our meaning system gets shattered and we have to construct a new one. But our culture does not value this process. We don't see any value in the shit. We want to flush it away. It takes courage to allow our faith and meaning to be dismantled. Despair can be a powerful path to the sacred and to a kind of illumination that doesn't come when we bypass the darkness.

— PSYCHOTHERAPIST MIRIAM GREENSPAN, IN AN INTERVIEW WITH PSYCHOTHERAPIST BARBARA PLATEK

Although it was midday, the sky darkened as in an eclipse, and as the column of birds thickened, their droppings fell like snow. For three days and nights the vast flock passed overhead, at a steady speed of 95 kilometres per hour (60 mph), undiminished and with no pause. At the last the very air smelled of pigeons, and their droppings had whitened the earth . . . At Audubon's reckoning, the flock that he experienced in the fall of 1813 consisted of an unimaginable 25 billion birds.

— TIM FLANNERY, ON PASSENGER PIGEONS IN JOHN JAMES AUDUBON'S WORK

Some might say there is too much shit in the world, in all the wrong places. And yet there are moments — watching a dung beetle at work, or being showered on by millions of rose-breasted birds — when one might see a strange beauty or certain poignancy in all this excrement.

Shit can be both awful and awe-inspiring, but it also has a purpose, and its beauty lies in its purpose. Psychotherapists Greenspan and Platek were not talking about the same shit that I have been talking about in this book, and perhaps I am pushing my luck to suggest that the path to enlightenment might be through understanding excrement. But there is a reason why the Egyptians deified the scarabs, other than their apparent propensity to deify everything from cats to onions. Scarabs (like onions and cats!) *are* special. Dung beetles speak to us of what we have lost, and what we must restore if we wish to live long and prosper. Based on an ecological and evolutionary transformation of our understanding of excrement from a "waste" to be "managed" to a necessity for life on Earth, we can restore shit to its rightful place in the biosphere, and in so doing discover both healing and meaning for ourselves.

Albert Einstein has been quoted as saying that we can't solve problems by using the same kind of thinking we used when we created them. This has become a kind of mantra for many public health workers, political activists, and environmentalists. Yet every time we (our species) begin to tackle a big problem — 9/11, a major oil spill in the Gulf of Mexico, loss of biodiversity, too much shit in the wrong places, obesity, starvation — we go back to exactly the same thinking that created the problem. Industrial technology created the problem of too much shit, and we think industrial technology will save us. New technology, important as it is, is only as good (in both senses of the word, morally and in terms of its effectiveness) as the context in which it is used and the challenges for which it is designed to respond. I would go so far as to say that developing the technology is easy; throw money at the engineers and they will invent something. Creating new technologies is not, primarily, the problem we have.

The science that underlies technology is what some academics in the natural and biomedical sciences call "hard science" and which is often the only kind of science considered "real" science. This results in natural and biomedical scientists making fools of themselves by preaching to the rest of the world about what *should* be done to respond to issues like excrement, when pretty much every intelligent person knows, based on good evidence, that such preaching is probably the least effective way to promote change. These "hard" scientists end up sounding silly because they clearly have no understanding at all of how and why people change. They imagine it is all due to ignorance, or cultural inertia, or that catch-all, "lack of political will." They also, ironically, act as if we (people) did not evolve through random mutations and selections inside the systems we are trying to understand. They often act as if we — or at least they — are objective, outside observers. The effect of taking this exclusively hard science, pseudo-objective approach is that anything having to do with the humanities — understanding why people behave the way we do, our understanding of what comprises knowledge, history, anthropology, ethics — is undervalued. It is often referred to as "soft science." Understanding how to use technology to create a sustainable society is considered "soft." I prefer the term "really difficult science," as proposed by geographer Barry Smit, for this work of co-creating a sustainable global narrative.

In part because of this lack of respect for the humanities, and in part because previous global narratives (Christianity, Islam, State Communism) have so often been catastrophically bad, the story many of us have told ourselves has focused on what we have seen to be the ideological "neutral" tale of technology and progress. We have deluded ourselves into believing that this is not a belief system, because it *uses* science to achieve its ends. But where this has led us, in the past century, is into a place where our stories have been constructed around single problems or built on narrow-minded academic disciplines. We have lived with the illusion that we can solve our problems one by one until they are all solved. If we look specifically at issues related to excrement, the international development literature is replete with tales of latrines being used to store food, or not used at all, because important social and ecological relationships have been ignored.

A story by one of my colleagues, Andres Sanchez, is typical.

In 1993 and 1994, Sanchez, an engineer-cum-anthropologist, visited the recently created Sierra Santa Marta, a Special Biosphere Reserve in Mexico. The reserve was home to more than 1,000 plant species, 400 bird species, and more than 1,000 other animal species, of which more than 150 species were listed as endangered. It was also the home of more than 60,000 Nahua and Zoque-Popoluca indigenous people. The reserve was also a major water source for an urban and petrochemical "corridor of Southern Veracruz."

Sanchez began, he says, with a rather traditional western approach to development research, exploring human behavior in relation to diarrhea and what preventive measures might be feasible based on a deeper understanding of relationships between people and their feces. Diarrhea in the area had been increasing annually, and in 1994, the cholera epidemic that was exploding throughout Latin America raised the anxiety of local citizens and government.

Although he eventually set aside this specific research in favor of a broader investigation of pre-Hispanic social and cultural beliefs about health and nature, Sanchez followed his original interest in shit through a series of more informal interactions with people in the area.

A review of the situation by the Ministry of Health found that the water source (a protected spring catchment) and distribution were safe, but that water was being contaminated by feces in the homes. Eighty percent of households had no water for washing or cooking in the home, nor sanitary facilities; those that did turned the water, as Sanchez recounted, into "fecal tea." People defecated in the open, women under the cover of darkness, and men in the privacy of their milpa (corn fields). Chickens, dogs, and pigs, roaming freely inside and outside homes, used feces as a main source of food in their scavenger diets, dashing after the fresh pickings left by children, who defecated in their home yards.

According to government officials, the problem was clearly one of lack of education and poor personal hygiene. Given this diagnosis, the government response appeared sensible. A school- and clinic-based community hygiene program was implemented, and the penning of animals was promoted. The doctor at the local health clinic gave a talk on sanitation and disease, and at the end of the talk provided a sack of cement and instructions to build a latrine. To attract women to the talk, the clinic gave a free kilogram of corn flour to all who attended. Since men were the ones who were culturally supposed to build things, and since many of them were uninterested in the need for a latrine, the cement was often left unused.

Pilar was a local woman who took a strong interest in the water, excrement, and health problems of the community. A Jesuit-trained community health worker deeply interested in communities' traditional stories, her diagnosis of the situation differed somewhat from that of the Ministry officials.

The population of the community had quadrupled over twenty years, and one group of people had cashed in on the coffee boom of the previous four years. Because of deforestation in the hills, the twenty-year-old water system was plagued by decreasing output in dry months. Responding to the periodic dry periods, the people who had become relatively wealthy from selling coffee used their extra cash to build home water storage tanks and flush toilets; sometimes they left their taps running, muddying the nearby streets, to the delight of free-ranging pigs.

Wastewater from the wealthy homes drained into a river upstream from an artesian well. The poor people in the community (who made up about 60% of the population) had to line up at public taps for their water. This was a task done by women and girls and often took much of the day, as the lines were long. Some simply did not have the time to wait: single mothers, women from households having someone with a chronic illness, or elder couples heading households with children whose parents had left home for work elsewhere or had died. The women and girls in these households then often had to do double shifts, working in the milpa and at home.

Eventually, the long lineups at the public taps caused these women to go to the artesian well near the river (downstream from the rich people's wastewater outlet). This well was often flooded, and contaminated, during heavy rainfalls. A water vendor who sold *agua fresca* (fresh water) outside the Sunday church service also drew his water from the contaminated well.

Because of the increases in population, the fields used for open-air defecation were moved farther away from the community. When men got sick with "the runs," they went to the field to relieve themselves, and came home to be cared for by their wives. Women could not afford this luxury. Sanchez heard stories of diarrhea-stricken women being beaten by men who came home after a long day of working at the milpa to find that the food was not prepared; or because there was town gossip of the wife seen going to the bush during the day, "probably to meet a lover." Women and girls, seeking privacy, often waited to defecate until nightfall; this also made them vulnerable to harassment and assault. On top of this, walking in the dark to the defecation fields, the women feared stepping on poisonous snakes that, at night, spread out on the roads trying to absorb the heat soaked up by the earth during the day.

The poorer households, often headed by single parents

or grandparents, or with a chronically ill family member, did not have the labor or time to build latrines. Pilar visited the municipal president and his wife, the head of social programs in the municipality, to put pressure on the local government to manage water with social equity and to lobby him to support a youth program to help the poorer sector build latrines.

After Pilar presented her version of events, she received a one-hour sermon from the president on democracy and equality of opportunity. He was unwilling to support such a program because, he argued, it was not democratic. Everyone had been offered the same opportunities to get a free latrine slab. Why should one sector now be singled out for special help?

Pilar returned home, determined to devise a new strategy. She began to build coalitions and partnerships in the community. She went over the head of the municipal president, lobbying NGOs to insert sanitation into the biosphere reserve management plan, a plan that had been developed and implemented by the state government. She lobbied her personal contacts and friends among the teachers and organized a nutrition and hygiene fair open to all, where she recruited volunteers for latrine-building groups. She kept the story in the local news networks.

When she returned to visit the municipal president and his wife, she came with a multi-stakeholder petition in hand for support of the latrine-building project. She organized a community meeting, where she helped give voice to dissatisfaction from community members about water waste, scarcity, and inequity. As the result of Pilar's persistence and insight, a community water supply and sanitation improvement program was inserted into the management plan for the biosphere reserve.

Pilar understood that dealing with excrement was not simply a matter of providing better toilets. It required addressing interconnections in an evolving context: a growing population, the new wealth of some that had led to the hoarding of safe water available to the community, snakes, community violence, poverty, inequality, and increased fecal contamination of the river — all complicated by the heavy inertia of culture and gender relations.

Sanchez explained to me that the story of this community in Mexico has been repeated in communities around the world. Excrement and water management are systemic issues, tightly related to inequalities of wealth, power, and gender relationships. Without addressing the social and ecological context, the wicked problem of excrement will remain intransigent.

We tend to think of such stories in relation to poor communities in "developing" countries. However, solutions to shit-related problems in industrialized countries are also, equally, embedded in stories, most based on carefully selected and biased evidence. What are some of these stories that have framed how we respond to excrement in our lives?

Several of the stories we live by are bounded, in good bourgeois fashion, by the assumption that our households are the center of the world. This is a kind of 1950s "family values" story, where all that matters is *our* clean house and *our* healthy children. This can be accomplished with a flush toilet, running water, and some enforced rules of etiquette. If we want to save water, we can use the motto I first encountered in my sister-in-law's bathroom in California more than a decade ago: if it's yellow, let it mellow. If it's brown, flush it down.

If the theme of our story concerns the invasion of a

neighbor's story into ours — say the offensive smell of our neighbor's commercial pig farm and the sight of his manure lagoons through our kitchen window — we can recommend certain feed additives and feeding regimes that change the smell of their shit and reduce the volume they produce. Or one of us can move somewhere else.

In the story we tell ourselves, we may we want to slow down the depletion of soils in Brazil, and reduce the shit-contamination of waters in Europe and North America. If we have a choice and can afford it, we can eat local, preferably organic, preferably from smaller mixed farms. We can eat less meat. Michael Pollan's advice is good: "Eat real food. Mostly plants. Not too much." We can keep fewer pets.

So far these stories are mostly about what is good for *us*, and only a little about what might be good for the planet that is our home.

Not all of our stories are completely selfish. If we want meat at a low price in the grocery store, we might also enlarge that story by saying that low prices are generally good for everyone, especially poor people. We might say that technological progress and creating wealth are the answer. In fact, we may not think there *is* too much shit in the wrong places. We might think that, in order for everyone on the planet to have a chicken in every pot, or pork cubes in every wok, or a steak on every grill, the amount of shit we have is the right amount, maybe even not quite enough. Soil depletion in some parts of the world and water contamination in others is simply the cost of having sufficient protein to adequately feed the world. This is a problem, we might even say, of social justice: it is everyone's *right* to have animal protein every day.

If the shit that comes out of these large farms is seen as a problem, well, we have a technological solution for that:

bio-digesters that produce electricity. Simply put, bio-gas is one of the outputs of systems called anaerobic digesters, or bio-digesters. These take organic matter (like manure, or animal or vegetable wastes), put it through a process of treatments that involve acid production and anaerobic (oxygen-hating) bacteria, and produce bio-gas (primarily methane, a powerful greenhouse gas), and a slurry. The gas is burned directly as a source of heat, or, more often, at least in larger commercial systems, to produce electricity. The slurry has been used as fertilizer, often after being composted or treated in another way. In some cases the water is separated and used for non-drinking purposes, like washing out barns. I'll have more to say about bio-digesters later.

If our story involves wanting to free people in large cities from the labors required to grow their own food, and to spend their time creating art and music, cars and computers, then we may find common cause with the economies-of-scale "feed the world" folks.

If the story we are interested in is about a diverse, resilient, sustainable, healthy world, a world that can anticipate and adapt to multiple possible futures, then we are back in the kind of narrative that interests me. In this story, everything — equity, enough food for everyone, how that food is produced and by whom and how it is distributed, the fate of excrement, art, ecological resilience, meaningful work — are all, simultaneously, important. In this world, ecological and social diversity are important, providing buffers against surprising changes in a world full of uncertainty: they provide resilience.[1]

1 Note that I am not only talking about diversity of organisms (animals, people, plants). It is the diversity of connections and feedbacks that are important for resilience.

But we also want to take advantage of some of the economies of scale, without going to the fantastic extremes preached by agricultural industrialists. Some of these economies, after all, free us to have more time to dance, write poetry, make art, and sing. We want a world where the tensions and arguments between local and global, past and future, change and memory, diversity and commonality, are real, ongoing, never-ending, and deeply rooted in our sense of home on Earth. To have this kind of a world, we must know shit.

In the previous chapter, I proposed the basics of a new way of thinking about, and responding to, the wicked problem of excrement and everything else. I suggested that a robust, post-normal science, rooted in a complex understanding of the world, will guide us in making our way to a healthy world. I have also suggested that there is unlikely to be a one-size-fits-all solution to the problem of excrement — that the solutions will be dependent on local ecology and culture, and justified by the narratives in which they are embedded.

Before we circle back to take a second look at some of the technologies available to us in moving toward this utopian vision, we should have a closer look at how one can elicit such stories. Since the mess we are in is largely the result of being held in thrall to a global techno-progress narrative, I am particularly interested in seeking an insurrection of marginalized, more complicated stories, such as the one shared by Andres Sanchez from Mexico.

Fortunately, a lot of people have been working on this knotty problem over the past few decades, and, in a kind of convergent evolution, most of the solutions these people are proposing have similar characteristics. The ecohealth approach, as I shall refer to it, has many names, comes in a variety of shapes and sizes, and has been "invented" by scholars and practitioners working on problems of public health, environmental management, conservation biology, economic and social change, and ecological resilience.

What you are getting here is the version that emerged from work in which I have personally been involved. Having set out a path by which we can walk from the wicked slough of excrement, I shall re-visit some of the many clever technologies that have been devised to help us with the details.

There is no paradigm for doing this; when the questions we ask are fundamentally, wickedly complex, as the questions I am asking about shit and sustainability are, then there is by definition no paradigm. Paradigms are important for sciences that are tightly constrained (physics, chemistry, communications, psychology). For the kinds of questions we want to address, we want to accommodate many different ways of knowing and types of knowledge and perspectives by microbiologists, energy engineers, economists, sanitary engineers, agronomists, public health workers, doctors, veterinarians, social scientists, social activists, old women, children, young men, Muslims, atheists, Catholics, anxious people in the streets, indigenous people, panarchy modelers, dancers, painters, poets, novelists, and political leaders. The expertise we need to solve the wicked problem of shit is collective. None of us has *the* answer.

In doing this collective work, we have to make up the rules together as we go along and keep testing them against the world we are investigating and transforming. It is the opposite of experts lecturing people; sermonizing by self-righteous industrialists and environmentalists reflects the same kind of thinking that created the mess we are in.

Is this possible?

For most of the past twenty years, I have worked as part of an international community of ecohealth scholars and practitioners. We are trying to figure out how to make the world a more equitable, healthy, happy place for people and all the other animals with whom we constantly co-create this amazing planet. Those, then, are the lenses through which I see the world. I see everything else — how many chickens or how much milk or how much shit we produce, how many cars, how many operas or novels or paintings — through those rose-colored lenses.

A decade ago, I worked closely with my late colleague James Kay, a systems design engineer, who was interested in sustainable management of the environment. He and his students created what came to be known as the "diamond diagram." The basic message of their work was that we need to bring together our best understanding of complex natural phenomena with our collective dreams and wishes to arrive at scenarios that accommodate both. Only then can we know what needs to be done, create programs to do those things, and figure out how to monitor whether things are moving in the direction we want.

At about the same time as Kay and his students and colleagues were doing their work, I was engaged with friends and colleagues in Kenya, Italy, Nepal, Peru, Colombia, Canada, and the U.S. to create a process for doing the kinds of things that Kay suggested needed doing. Today, there are ecohealth networks and communities of practice in Canada, Latin America, Africa, and Asia. These communities include practitioners and scholars of all types, from economists, ecologists, engineers, doctors, veterinarians, and communications specialists to farmers and community organizers. Globally, the International Association for Ecology and Health has brought many of these people together. Our meetings are confusing, eclectic, exciting, frustrating, and inspiring.

The accommodation of different perspectives is rarely easy, especially for a scientist such as myself who is used to being "right." One particular instance brought this home to me. In October 2010, I participated in a workshop delving into ecosystem services for poverty alleviation, in particular the relationship between biodiversity and indigenous health in the Amazon and the Yungas, the eastern slopes of the Andes. The workshop included physicians, anthropologists, epidemiologists, veterinarians, teachers, ecologists, and naturalists. We had indigenous leaders, Argentinians, Canadians, Brazilians, Brits, Peruvians, Colombians. Over a week, we argued, drew pictures, showed maps, yelled at each other, cried, laughed, got drunk, made lists and organizational diagrams, came back to the table, checked with our social networks on Skype, and drew up a research plan. What we were forging was an uneasy sense of common vision — One Health, global solidarity, mutual respect, finding interwoven narrative threads. In the midst of this, one of the indigenous leaders said that the traditional way of dealing with enemies, with people with whom they disagreed radically over important issues, was to kill them. She felt like killing us. The fact that our discussions elicited this response, and that she did not kill us, was reassuring for me. It meant that we were dealing with important issues, that the perspectives we brought to the table were, in fact, substantially different, and that there was hope that we could, if not resolve our differences, at least not kill each other because of them.

One formulation of the ecosystem approach to health, the one I have been mostly closely associated with, is called an Adaptive Methodology for Ecosystem Sustainability and Health, or AMESH. In simplified form, it boils down to a series of identifiable "steps," although the steps are usually not as orderly as my presentation of it suggests, and sometimes we end up walking in circles, revisiting the same steps a few times, or jumping to the end before returning to earlier steps and then moving on.

1. In the first place, there is a presenting "complaint" or issue or problematic situation: Why are we here? Who invited us? To the best of our knowledge, how did this situation come to be (the collective histories as generally, publically, accepted)?
2. Who are the participants — often called stakeholders — in this situation? What do they care about? What rules (official or unofficial) govern their behavior and decisions? What are these rules based on? Gender? Race? Wealth? Caste? Class? Aboriginal status? Are there alliances and conflicts between the different groups? Where do non-human species fit into this?
3. What are the histories and narratives told by the players about how the situation came to be and their role(s) in it?
4. What is our best, systemic, scientific understanding of this complex situation?
5. What is our best understanding of the social and cultural issues that need to be addressed?
6. How are 4 and 5 related? How do they "feed" off each other and constrain each other?
7. What are the scenarios, visions, or narratives

that people most connect with? What are the things that people agree on? On which things will we likely not achieve consensus? How will different actions influence these collateral issues? How do issues of equitability in power relationships reflected in gender, age, race, economic status, and the like impact these narratives? This is also the point in the process where I tell everyone that, American dreams notwithstanding, you can't be anything you want to be. We all have limits, as do the natural systems of which we are a part. Realistically, then, what do we, collectively, wish to do? What story do we want our grandchildren to tell about us?

8. What kind of governance structure and course of action will enable us to move ahead to implement those visions, and to move toward the goals, on which we have agreed? What kind of a monitoring system will enable us to determine whether we are achieving what we set out to?
9. Implement. Monitor. Adjust. Learn. Re-Start.

Although AMESH emerged from projects and research in several countries, my experience in Nepal was particularly instructive. I went to Nepal in 1991 to investigate a human disease (called hydatid disease, or cystic echinococcosis) related to dog shit.[2] *Echinococcus* tapeworms

2 This work is described in more detail in my book on the ecological and evolution of diseases people share with other animals, *The Chickens Fight Back* (Vancouver: Greystone Press, 2007). For those who wish a deeper scholarly analysis of this, see *The Ecosystem Approach: Complexity, Uncertainty and Managing for Sustainability* (New York: Columbia University Press, 2008) and *Ecosystem Sustainability and Health* (Cambridge: Cambridge University Press, 2004).

reproduce sexually in the intestines of canids (dogs, wolves, coyotes, foxes) the world over. The gravid tapeworm is excreted, but can only complete its cycle if it is eaten by another species. In that other species, which is usually some form of ruminant (sheep, cow, water buffalo, moose), the tapeworm forms cysts, which are like slow-growing tumors full of tiny "proto"-tapeworms. When the second animal dies, and a canid eats the cyst, the tapeworms can mature and have sex in the canid intestine, and the life cycle is completed. In many parts of the world, this life cycle evolved to take advantage of the fact that people domesticated both sheep and dogs, and generally took pretty good care of both of them. People get this tapeworm cyst — again as a slowly growing tumor full of baby tapeworms — by accidentally eating dog poop. The habit of keeping sheep and sleeping with your sheep dog to keep warm perpetuated the parasite cycle over many millennia. North American versions of this life cycle involve moose and dogs, or wolves and caribou, or foxes and voles, as well as sheep and dogs.

In Nepal, the cysts were arriving in goats and sheep from the high plains of Tibet, and water buffalo from the hot plains in the south of Nepal and the north of India. When I entered the picture, invited by Nepalese colleagues to help them investigate this parasite and how to control it, the goats and buffalo were slaughtered at open sites along the riverbanks of the Bishnumati River in Kathmandu.

Initially, I saw this as a pretty straightforward problem: people were exposed to dog poop. How could we stop that? As I worked with my Nepalese colleagues throughout the 1990s, we realized that the problem involved, at a minimum, providing meat for the tourist industry, a mainstay of the economy; importing animals to maintain the supply; slaughtering those animals in the open air along the riverbank; letting organic garbage pile up in the street (saving disposal money, but also providing an extra source of nourishment for the street dogs); tolerating and even encouraging free-running packs of dogs (which provided community policing functions at night); and allowing much-loved pet dogs to defecate in the house.

After several years of trying, unsuccessfully, to "solve" the problem of people getting hydatid disease through basic research and public lecturing, we decided to change our thinking and our strategies. Going well beyond the simple problem of dog excrement to tackle the more wicked challenge of "everything at once" we worked together with butchers, slaughtering workers, street sweepers, garbage collectors, local politicians, shop-keepers, social activists, and researchers (veterinarians, parasitologists, anthropologists). Not long afterward, the community completely restructured itself, changed leadership in the Butchers' Association, created closed-in slaughter facilities, moved holding pens for buffaloes to fields outside the city, started composting the offal and feces of slaughtered animals, stabilized the riverbanks with parks and grasses, built public toilets, and looked at how to create better garbage collection (which may, in the long run, require daycare and schooling for the children of the young mothers who sweep up the garbage). The problems of environmental contamination with excrement and the parasite infection were addressed by not seeing them as isolated issues. The shit was deeply embedded in what Douglas Adams would have called "life, the universe, and everything."

It also seemed to me that this community-level reorganization in Nepal probably would not have happened if the

Berlin Wall had not come down, because the activism that led to all these changes was part of the global movement for democratization in the 1990s. Complexity theorist John Casti has called this the "social mood," and argues for its critical importance in how events in the world unfold.

All of these activities and events fit in with a holonocratic, panarchic view of the world, where behavior at one scale (personal, household, community, watershed, region) influences, and is influenced by, behavior at scales larger and smaller. What has impressed me as well is that the communities we worked with continue to take on new projects, and have adapted and responded to new challenges in the midst of a decade of political unrest and civil war. A good deal of the excitement and energy came from recognizing that there was such a diversity of things that could be done. Everybody had (has) a role to play. This resilience of the local community is hopeful, since it means local resilience can survive larger scale collapse — that is, the collapse of national government in this case — and serve as a source of renewal and inspiration for people who are trying to renew the larger system.

When big industrial, globalized livestock-rearing systems (like centralized economies and political systems of all sorts) collapse, renewal will be possible if smaller-scale, integrated, inventive, diverse animal and shit-management systems are already in place.

This story about Kathmandu is not perfect, nor is it an isolated event. It represents a new way of thinking about shit, life, and everything, and, like all new thinking, is inventive, exciting, subject to local power politics, and vulnerable to being overwhelmed by larger-scale events (climate change, pandemics, population explosions, migration to cities). Nevertheless, if we apply AMESH-style thinking at any scale of endeavor, keeping panarchy and holonocracy as our mental orientation, we can begin to resolve the global and local shit problems.

In every situation where I have worked using these orientations, people get excited when they see that lots of different kinds of knowledge are valued, and that the solutions will be collective, rather than being imposed by some outside expert.

Technology, as I have said, is important. Nevertheless I hesitated to include a lot of technological responses to the wicked problem of shit for two reasons. Firstly, having spent several decades as a scholarly researcher, I am well aware that most applied research is driven by funding. If money is made available, viable technological solutions to specific problems will soon appear. Secondly, a new global awareness of excrement, primarily as a public health problem, has elicited the money that will drive that research.

For instance, the Bill & Melinda Gates Foundation's "Reinvent the Toilet" Challenge in 2011 resulted, within a year, in prototype toilets that produced fertilizer, electricity, and clean water. As I would have predicted, however, based on several decades of evidence from "soft" science, it is not clear how, or when, such technologies will be adopted. It is not simply a matter of "transferring" technologies, as some naïve development "experts" used to think. It is a matter of working with people where they live to co-create a narrative within which those technologies make sense. That is difficult science. Having said that, I would be remiss if I did not at least discuss some of the basic technologies that are currently available. Some of them are quite simple and have been around a long time. Others are very new.

Probably the most widely accepted use of both human

and animal manure is as fertilizer, which would also seem to be the closest to mimicking natural ecological cycles. Indeed, it is when excrement is *not* treated as a powerful fertilizer that we run into such major problems as nitrate pollution of drinking water and toxic algae blooms in the oceans. I have already explored this aspect of manure at length, but it is worth underlining here that fertilizer is not only the past of shit. Excrement is likely to continue to play an important role as fertilizer in the future, along with its use as a source of energy, which I shall return to later. Both of these uses are driven by increases in fossil fuel prices, as well as by popular demand for organically produced food.

Although not as highly valued as it was in the nineteenth century, guano has been regaining some of its former "luster." According to some news reports, Quechua workers gathering guano in the islands off the Peruvian coast have been earning three times what they could earn back in their highland homeland. The guano is shipped back to mainland Peru to support organic farming. In a move that could herald a new frontier in species protection, the Peruvian government stationed armed guards at each of the more than twenty islands to keep away anyone who threatened the guano-producing Peruvian boobies and Guanay cormorants. Harvesting rotates among the islands, so as to minimize human intrusions. The government is also controlling commercial fishing in the area of the islands, in order to conserve the fish-eating guano-producer populations of birds.

The revival of the manure-as-valuable resource will accelerate as we begin to treat livestock agriculture seriously (recognizing it as the ecological manipulation that it is) and as we take the long, inexorable slide down the far side of "peak oil" and away from petroleum-based chemical fertilizers as the price of oil increases. There are signs that this trend is already occurring.

The oldest technology for managing feces is composting. Composting is a special case of what in nature would be called decomposition; in natural systems, organic matter is broken down and becomes humus. Although all organic matter decomposes through normal microbial action, managed composting is quite different than simply dumping manure or dead animals into a hole in the ground and waiting for them to rot.

As part of an ecosystem health course for veterinary students, we had them consider what to do with dead chickens in the face of an avian influenza epidemic; it would not be wise to load them up in trucks and take them to a landfill, dripping blood and viruses along the way. So what should a small- to medium-sized farmer do? We had the students toss some dead chickens onto a pile, cover them with straw, add some fuel, and set fire to them. We had them throw another few into a hole in the ground and cover them up with dirt. In another hole, they layered chickens, dirt, and straw, and left aeration tubes in place; this was the compost hole. Long-stemmed thermometers were inserted into the two holes. Burning the chickens took up a lot of wood and straw, and set up a terrific, exciting blaze that could be dangerous to anyone in the area; you wouldn't want to try that during a drought, or in the mid-summer heat. After forty-eight hours, the composted chickens had heated up and were rapidly being reclaimed by soil insects and microbes. The buried chickens were pretty much as we had left them; the bacteria needed to accelerate the process need oxygen and more carbon than the carcasses provided.

Researchers have demonstrated that a well-constructed

compost pile, with the appropriate mix of carbon, nitrogen, and oxygen (to promote the appropriate bacteria), can reach temperatures of 54–66°C (130–150°F), which is enough to kill avian influenza viruses. After a couple of weeks, usable soil and perhaps a few feathers and bones are all that is left.

When my family composted one of our house cats (who was hit by a car), she was returned to our garden as excellent soil, and a few tiny bones, and was resurrected to us as flowers and vegetables. If manure is used instead of dead chickens, the same process occurs. It's especially effective if the manure is mixed with bedding materials, to give a better balance of nitrogen from the shit and carbon from the wood or straw. The same process could be used to render cat and dog shit reusable, but public health officials are somewhat nervous that people not familiar with composting would simply bury the feces, and hence provide opportunities for a variety of parasites to leach into the surrounding soil and water. Composting can be done in the ground or, if larger volumes are available, in long windrows or piles on concrete slabs that are turned periodically to aerate them.

On a larger scale, the return of shit as a valuable product brings with it some serious challenges, especially in the context of globalization of animal feed supplies. Where the shit is produced is usually considered, but where the feed inputs to the animals and people that produce the shit comes from is usually ignored.

Industrialized countries have promoted the spreading of biosolids on farmlands. New technologies for treating excrement has made the transmission of pathogenic bacteria less likely. In Ontario, Canada, where I live, standard sewage treatment removes about 90% of the pathogenic bacteria.

The major problem with biosolids is not that they spread infectious disease but that they often carry heavy metals such as lead, mercury, and cadmium, which do not die off or wither away even under the brightest, hottest sun. As the result of new technologies and regulations, in many cases the levels of these metals is much lower in biosolids produced by industrialized countries now than it was a few decades ago. But over a few decades, continuous applications on the same lands can lead to dangerous levels because of bioaccumulation. The implications of these higher levels of chemical elements is not yet clear.

If we mimic natural systems, then we would scatter smaller amounts of (preferably composted) feces over wider areas, the amount and placement being determined by climate, soil type, vegetation, slope, and the like. This would imply that we should back off the biggest-is-best model and think more in terms of farms sized to fit their surrounding ecosystems and social systems, with excrement-recycling systems tailored to fit the context.

Closing the cycles of energy and nutrients to create resilient ecosystems is more easily done locally than at industrial scales. Using bio-digesters and composting provide some interesting options. Integrating fish ponds into a mixed farm is another possibility. The addition of human or animal excreta to fish ponds (usually growing carp) has been recorded throughout Asia (particularly China), Egypt, and Europe for many centuries. More recently, agronomists have run experiments using animal manure to fertilize tilapia-growing ponds in Africa. This practice, by enriching the water with nutrients, promotes the growth of bacteria, algae, and zooplankton and results in good quality fish protein. It is thus a variation of the use of manure as fertilizer.

As a member of a class of scientists (epidemiologists) who have encouraged people to eat more fish (for heart health), and at the same time being concerned about both water contamination *and* the collapse of the global open-water fisheries, I am attracted to the notion of growing fish in shit-infested waters. Nitrates and phosphates from the manure are kept in the food chain, and fish help aerate the water, which discourages the growth of disease-causing bacteria. The use of animal feces to feed fish is one of those delightful eco-friendly farmer stories and scenarios much celebrated in the 1970s, until they were bulldozed by urban demands, misconceptions about our apparent "freedom" from ecological constraints, and diseases such as avian influenza that gave multi-species farming a bad name.

If there are concerns about the chickens carrying human pathogens such as *Campylobacter* or *Salmonella*, composting or otherwise treating excrement to kill off pathogens before it is fed to the fish, clearing away vegetation at the edges of the ponds to discourage the growth of snails (which can carry the parasites that cause schistosomiasis, a disease that damages internal organs in humans), and "flushing" the ponds with clear water a few weeks before harvesting the fish will help to result in a cleaner, healthier food. As with any option we have on the list of "what to do with this shit," this one has to be carefully managed.

While manure's most prominent role historically has been as fertilizer, this has been changing with increasing energy costs. The use of excrement for fuel is no longer restricted to the burning of cow pats in India and Nepal. Given that energy is probably the single greatest limiting factor for urban and industrial development, it is not surprising that developments in finding new energy-related uses for manure are perhaps the furthest along. Most of the technological innovations related to energy production from manure are related to the generation of bio-gas.

Bio-gas generation requires slightly more advanced technology than simple composting, but also offers a few more advantages, especially for up-scaling to larger farms or dense urban populations of people and dogs. In international development circles, the use of bio-digesters to produce bio-gas has a long history in places where there is both a shortage of fossil fuels and an excess of manure. The process of bio-digesting has been the subject of many books and development projects, as it seems to offer a rare win-win situation. Rose George, in her book *The Big Necessity*, devotes a considerable number of pages to the Chinese efforts in this regard. According to *The People's Daily*, the Chinese claim to have 748 large- and medium-sized digesters that handle 20 million metric tons of human sewage annually and produce 200 million cubic meters of methane gas.

Although bio-gas production through anaerobic digestion in a bio-digester could be carried out at one of a number of temperatures, one needs to pick one temperature and stick with it so the bacteria that are doing the work feel "comfortable." All bacteria have a range within which they best grow, multiply, and use resources. *Listeria*, for instance, prefer refrigerator temperatures to grow and multiply. *Salmonella* prefer temperatures that are closer to those of a mammalian body. To take advantage of different kinds of bacteria, and variations in ambient temperature, there are thermophilic (50°C–60°C), mesophilic (35°C–40°C), and psychrophilic (15°C–25°C) digesters. The communities of bacteria that survive naturally in thermophilic digesters multiply and work faster at higher temperatures; they can process at least small amounts of material in three to five days. However, the bacteria that thrive in these hotter digesters are also quite sensitive to temperature and pH fluctuations, and these digesters need to be more carefully managed than those that work at cooler temperatures.

The mesophilic digesters take longer than thermophilic ones (fifteen to twenty days), aren't as efficient in killing pathogenic bacteria, and produce less gas. Because of the much slower bacterial action at lower temperatures, the psychrophylic take even longer and are less efficient at breaking down the organic matter. The choice of one over the other depends on ambient temperature (low tropics versus snowy mountains, for instance), what kind of material is to be processed (manure, dead animals, straw, vegetable waste), and how much (a farm's worth, a whole village's).

The cooler digesters may not achieve sufficient pathogen kill for the slurry output to be safely put on fields where foods are grown for human consumption. That means the cooler plants' output may still need to be composted (which, unlike the bio-digesters, uses aerobic bacteria) to kill the pathogens. In any case, bio-digesters have been designed for just about every size and location in India and Nepal, and are widely used in those countries as a way to reduce dependence on burning wood or coal, and to prevent respiratory disease, which commonly occurs in women who cook over traditional smoky wood fires.

Engineers working for Hewlett-Packard have suggested that, with big enough input, electricity from biogas produced from cattle manure could be used to run computer centers. Similar systems have been described (and are in use) for an 1,800-head dairy farm in Maine and a 9,000-head swine farm in North Carolina. In North

Carolina, the system, created with input from Google Inc., was used to claim carbon offsets. According to the farmer, the system reduces waste emissions, improves the health of the pigs, and creates a fertiziler he uses to grow wheat, corn, and beans. Some are now reporting that Google and Apple are competing to tap into the fecal energy produced by North Carolina's large hog farms.

While bio-digesters in North America are usually related to livestock, this is not the case elsewhere in the world, where human crowding under unsanitary conditions is more common than livestock crowding. Dozens of community "biocenters," large enough to accommodate a thousand people per day, have been built in crowded slums around Nairobi, Kenya. These centers contain hot showers and, sometimes, offices and other businesses above them. The centers, including the hot showers, are fueled by human excrement. The bathrooms were built over bio-digesters that decomposed human manure and urine to produce methane gas, which was supplied to kitchens in the surrounding areas. This solution not only provided fuel for cooking in these unserviced slums, but also eliminated the need for "flying toilets." These plastic bags full of human excrement, thrown into the streets at night, were reminiscent of waste disposal habits in sixteenth-century London.

Post–civil war Rwanda has become a world leader in imaginative use of excrement. Researchers and development workers identified organic waste, including excrement, from prisons and secondary schools as an important health hazard to areas surrounding these institutions. At the same time, both prisons and schools were contributing to deforestation because of their demands for fuelwood for cooking. One of my Rwandan colleagues wrote to me that "the Kigali Institute of Science, Technology, and Management (KIST) developed and installed large-scale bio-gas plants in prisons and secondary schools. Each prison was supplied with a linked series of underground bio-gas digesters, in which the waste decomposes to produce bio-gas. After this treatment, the bio-effluent is safe to be used as fertilizer for production of crops and fuelwood. KIST staff manage the construction of the bio-gas plants, and provide on-the-job training to both civilian technicians and prisoners. The first prison bio-gas plant started operation in 2001, and by 2011 plants were in operation in ten prisons. The largest has a series of twelve individual digesters." According to this colleague, many homes are now installing similar digesters, but on a smaller scale.

The example from Nairobi is as close to a win-win situation as one might hope for; the use of prisoners' waste to produce power for jails in places like Rwanda makes me just a little nervous. The human-waste bio-digesters in Rwanda produce about half the power needed in dozens of the country's overcrowded jails. On the plus side, using this "natural gas" helps spare forests from being cleared for fuel in Rwanda's already denuded countryside and keeps the manure out of the waterways used for drinking. It also provides excellent, odorless fertilizer for the prison gardens, which produce food for the inmates. On the downside, this constant need for the raw materials used to create the power would seem to require that the authorities keep the prisons full in order to keep them running, regardless of the crime situation. At this point, this does not seem to be an issue. And I suppose that the jails, if they are emptied, won't need power sources. And one could do a lot worse than having an energy system for schools dependent on keeping the classrooms full.

Like the Rwandan jail bio-digesters, building large bio-digesters (such as those suggested by Hewlett-Packard and those implemented on large North Carolina hog farms) raises some serious dilemmas. On the one hand, it would be an excellent use of the vast amounts of manure produced by large livestock operations, helping to reduce the environmental impact of what is clearly a dirty business. On the other hand, once built, such plants would *require* these large inputs of manure or other organic matter. This choice, then, closes other options. Do we want to have our energy dependent on large pig farms? What are the other environmental and social impacts of those farms?

In any nested social-ecological system, "scaling up" almost always means a loss of adaptability and an increased vulnerability to large-scale failure, especially in the face of increased instability in political, economic, and climate systems. A breakdown in a small bio-digester is a manageable problem. A breakdown in a large one can have serious, cascading effects through the whole system. Perhaps a cooperative system involving multiple smaller farms and digesters linked together might offer the best of both worlds.

Globally, a decrease in human populations and a decrease (probably a drastic one) in intensive livestock rearing are in the best interests of the seventh generation into the future. Having made such a global statement, however, I would add that, even as North Americans and Europeans and wealthy people everywhere should decrease meat consumption and its attendant shit production, I think that very poor people throughout the tropics should have the option to eat more meat. Nutrition researchers have demonstrated that kids learn better in school if they have some animal protein in the diet. But that doesn't require large-scale intensive livestock rearing.

Ecologically and socially (that is, in terms of our best understanding of complex, adaptive, self-organizing social-ecological systems within which all life is embedded), it makes much more sense to have a lot of smaller livestock operations well integrated into local farming activities to simultaneously facilitate more effective manure management, stronger rural communities, and more diverse landscapes.

The issue of the impacts of all technologies, not only bio-digesters, therefore raises the broader questions of economies of scale. The farms and industries required to achieve the kinds of cost savings that make such economies attractive take up large pieces of land, and promote populations of animals that are genetically similar. These enterprises are thus associated with the loss of biodiversity and adaptability in the social and natural landscapes where they are situated.

I am not so naïve to think that large-scale livestock operations will disappear overnight, or ever. Some large bio-digester plants make sense. But I propose that a variation in size would be more appropriate to meet the multiple goals of public health and ecological sustainability — some large plants and many medium or smaller ones. Globally, there is a lot of evidence that this variability in size is happening, and that the smaller, faster scales of innovation in the panarchy will continue to thrive.

Historically, not all urban centers have come up with the same solutions to fecal pollution. Over many centuries, for instance, the Yemenis developed elaborate systems to separate urine and excreta even in multi-story buildings. Urine passed from toilets along a channel to the outside wall of the building, where it evaporated in the hot dry climate, but feces were collected from toilets via vertical drop shafts. The feces were then sun-dried and burnt as fuel. This sanitation system required very little water, an advantage in the dry desert environment. The "modernization" of Yemen, which included the introduction of flush toilets, has been associated with water shortages and falling water tables in the capital city of Sana'a.

Sweden is one of the world leaders in energy produced from biological materials. Since 2005, the world's first bio-gas–powered train, the Amanda, has been running the 120 kilometers between the cities of Linköping (Sweden's fifth-largest city) and Västervik. In Västervik, the source of the bio-gas is the local sewage treatment plant. In Linköping, where buses and garbage trucks are fueled by bio-gas that is available at gas stations, the bio-gas is produced using the wastes from a local abattoir.

The use of cattle manure and other waste to produce power is more widespread in the world and less problematic, from a public health point of view, than the use of human excrement. In Nepal and India, where cow dung provides energy for more than a million people, it may be the "once and future" king of energy. If a cow defecates in the street, someone most certainly is watching, and will scoop up the patty and slap it onto a wall to dry. Once dried, it makes a reasonable fuel.

Cow dung has about the same heat-producing value as wood (but both have less than half of what kerosene provides). Llama dung has about the same value as that of cattle. Globally, 40 to 50% of the 150 (dry matter) metric tons of cow dung used annually for fuel is burned in India. While the traditionally flat patties are still common, the efficiency of energy conversion can apparently be raised from 10 to 60% by putting the dung through an anaerobic bio-digester. Since the killing of cows is forbidden by Hindus, the Indian subcontinent is likely to maintain a large cattle population for many generations to come. In this context, it would seem that finding new jobs for cows in India (such as eating garbage, producing fuel, and protecting forests) would be easier than changing the respectful Hindu attitudes toward the gentle, stubborn, smugly self-righteous beasts.

The global translocation of nutrients through animal feeds and human foods is a classic, wicked problem of complexity. The logic of nutrient cycles would suggest that one should send the shit, produced by the animals and people that eat the food, back to the countries from whence the foods came. But it also makes some sense, if we are actively managing the biosphere, to ship the shit to places where it might be useful for other reasons.

For instance, in the 1990s, Dutch entrepreneurs planned to export some 7 million metric tons of cow manure to India to be used as fuel. It seemed like an elegant solution to the "too much shit" problem the Netherlands were facing. Given the high levels of antibiotic use in Europe, however, the Indians feared that the dung might bring with it antibacterial-resistant bacteria. Bacteria share genetic material, including genes that code for resistance to antimicrobial drugs, without regard to any religious rules about appropriate breeding behavior. These genes can be transferred from normal gut bacteria that don't usually cause problems to serious pathogens such as *Salmonella* and *Campylobacter*. Researchers have already determined that antibiotic-resistant bacteria have spread from the shit of people to mountain gorillas in Uganda, and can spread from human wastes to foraging gulls. The gulls can spread these bacteria, and their resistance genes, wherever they fly.

The program proposed by the Netherlands could thus

solve a manure problem in the Netherlands and a fuel energy problem in India, but create problems for treating animal and human diseases by increasing antibiotic resistance. In the end, the project was dropped. Necessity being the mother of at least some inventiveness, in 2009 a report from Amsterdam announced that a new bio-gas plant had been opened near Leeuwarden that would use cow dung, grass, and food industry "waste" to generate sufficient heat for more than a thousand homes.

While global trade in livestock excrement and large-scale bio-digesters tend to grab headlines, often the best solutions are more nuanced and tailored to local conditions. If people in India can generate fuel from cow pies, feedlot owners can use cow shit to run computers, and Swedish engineers can help manage urban waste by using it to run buses, what do millions of North American city-dwellers have to offer? What about all those dog owners running after their pets with plastic disposal bags? Can we recruit them into the power grid?

Dogs and cats also produce huge amounts of organic urban waste, much of it going to landfills or leaching into waterways. A dog park in Cambridge, Massachusetts, provides a kind of pilot project for what can be done. Dog owners in the park collect the poop in biodegradable bags, and deposit it in the "Park Spark" project. This project, the work of artist Matt Mazzotta, digests the poop and produces sufficient methane to run a street lamp. Norcal Waste, a garbage company based in San Francisco (a city that an estimated 120,000 dogs call home) has proposed that dog waste be collected and converted to biofuel. The estimated 9 million metric tons of shit produced by American dogs and cats (which are fed diets richer than those of most people in the world) may yet contribute to that country's energy self-sufficiency, although I have yet to see it mentioned in the national energy plans.

While I was writing an early draft of this book, the workers for the city of Toronto who collect garbage were on strike. Dog owners lamented on the radio that they not only had to pick up the dog turds (which by law they must do), but now they had to take them home and do something with them other than put them into the trash. What hardship! Some have recounted harrowing and stinky tales of burying dog poop in the backyard. Why didn't they mix it with grass cuttings and kitchen waste and compost it? The soil in their backyards would have been much improved, without stink, and public health and well-being would have been better served. Apartment and condominium owners (where dog owners do not have a yard) could band together and either compost or start a biofuel business, as proposed by Norcal in San Francisco. I can see the bumper sticker already: *My car drives on dogshit!*

On a slightly different tack, in the wake of the disaster in the Gulf of Mexico, and as oil supplies dwindle, Yuanhui Zhang, an agricultural engineering professor at the University of Illinois, may one day be seen as a kind of cutting-edge hero. He has apparently been able to convert two liters of pig manure into a quarter of a liter of oil through a thermochemical process. Not much, but it is a start.

As the cost of oil has increased, we've seen increased interest in the use of manure as fertilizer and also as feed. Since the microbes in cattle rumens can take nitrogen sources and combine them with carbohydrates to manufacture protein-rich feeds, chicken manure can serve as an inexpensive substitute for grain and protein supplements in cattle feeds. This has been done in various parts of the world with, it appears, mixed success. While the shit serves as a good protein substitute, in one study in Israel, some of the chicken feces were found to have high estrogen levels, which interfered with normal development in young cattle. Any time anything is recycled, especially in intensive systems where problems can be magnified, a good quality control system is important.

One possible alternative to the direct use of manure as feed, if that is deemed too risky, is to raise flies on the manure, and then to dry and process their larvae, which are an astonishing 40% protein. This feed can then replace corn or soybeans in the diets of cattle, chickens, pigs, and fish — or ducks. Butchers in the neighborhoods of Kathmandu I described earlier, where people were suffering from hydatid disease, composted manure and offal in open rows. Others in the community then raised ducks; the ducks fed on the bugs that transformed the "waste" into food. This is relatively straightforward when done on a small to medium scale. The technology to achieve this on a large commercial scale will be mixture of the mundane (growing flies) and the high tech (creating harvesting chambers, post-harvest treatments of the flies, and the like).

Besides the usual options to use manure as fertilizer and energy, a variety of novel uses have been proposed, from jewelry to novelty gifts. At present, most of these uses are restricted to "niche" markets. Although I have my doubts about the commercial viability of the meat created from sewage sludge that I mentioned in an earlier chapter, some of these "novel" uses may, in particular contexts and in the interests of diversity, become important.

Earlier I mentioned the use of civet excretions and ambergris from sperm whales in perfume, but excretions out of, or from near, the anuses of animals have also served a variety of other purposes.

Many termites use feces (as well as soil and wood) in the construction of their mounds. As well, some birds (gannets, kittiwakes, South American oilbirds, the wonderfully strutting secretary birds) use their own and others use that of other animals to line their nests. Not surprisingly, then, people have experimented with similar uses of excrement. The Maasai, Dinka, and Nuer tribes of Africa use cattle dung as a type of mortar in the construction of their earth-brick houses, and in Belarus cow dung has been used to compact the thatch on cottage roofs. In India, dung is mixed with mud and used as flooring in rural areas. The gut bacteria and residual, undigested fibers present in the dung create a smooth, durable floor that produces less dust than mud alone. In Indonesia, a company called EcoFaeBrick produces bricks from cow dung that are said to be 20% lighter and 20% stronger than clay bricks, and provides increased income for farmers.

Fossil fuels are used for making a great many plastic-related compounds that have become essential in hospitals (disposable needles, intravenous bags), common in cafeterias and classrooms (plastic chairs), and desirable for backpackers and hikers who prefer to wear synthetic jackets made from fossil fuels rather than animal skins. Almost 5% of oil consumed in North America is used for making plastic. At least some of those plastics can be made from excrement. Micromidas, a California company, has developed a commercial process for creating plastic from wastewater and sewage sludge.

In another twist, in 2006, Japanese researcher Mayu Yamamoto of the International Medical Center of Japan was presented with the Ig Nobel Prize in Chemistry for extracting vanilla from cow dung. The one-hour heating and pressuring process costs less than half of what it costs to extract vanilla from vanilla beans. Most synthetic vanillin is made from petrochemicals; however, the lignin used to make vanillin is also present in the fecal matter of grass-eating animals (primarily ruminants), and given the amount of livestock feces in the world and the fact that we are past peak oil, this form of vanilla would seem to be an interesting option. But try not to think about it as you make your pudding.

As I mentioned earlier, elephants only digest about 40% of what they eat, and they put out 100 kilograms (220 pounds) of feces per day. The 60% of undigested, or partly digested, material can of course be reused, and is. Warthogs eat elephant dung, as will elephants themselves (if they are really hungry). However, elephant dung can be used to make paper, as can kangaroo dung, which is plentiful where elephant dung is not. Elephant dung paper, unlike similar products from other animals, has become something of a cause célèbre among those who consider themselves to be environmentally friendly. This may be because elephants are (rightly, in my view) not considered appropriate animals for the barbecue, or perhaps because our collective efforts to save the rainforests have inadvertently put many working elephants in Southeast Asia on the unemployment rolls.

On a visit to the Thai Elephant Conservation Center in the dry, rolling hills south of Chiang Mai in northern Thailand to check on a project for Veterinarians without Borders/Vétérinaires sans Frontières, I could not leave without exploring what happened to the elephant dung, which, as one might imagine in an elephant retreat center, was plentiful. A large, round, yellow sign on the wall declared (exactly as it appeared):

How we help the elephants?

1. The dung is washed and boiled to kill any bacteria. Not as bad as it sounds, elephant dung doesn't smell!
2. A mild bleach is added that will not harm the environment.
3. The dung is spun for up to 3 hours to cut the fibres. Color is also added.
4. 300g balls are rolled — enough to distribute over a frame. Mulberry paper is made in a similar way.
5. The frames dry naturally in the sun. No 2 sheets of Paper are identical as the coarseness is dependant on the elephants diet!
6. The paper is gently sandent [*sic*] to make it smooth enough to write on. A range of products are made by local people.

One elephant is said to supply sufficient dung for 115 sheets of paper every day. The estimated 500,000 elephants in the world could thus provide enough poop, daily, for more than 50 million sheets of paper. As with any technology, "scaling up" and diverting elephant feces to produce paper on a large scale would remove that manure from the ecosystems where the elephants live, hence remove that as a source of food for other animals, including dung beetles, and probably result in a general deterioration of those ecosystems. Nevertheless, it is attractive as a cottage industry for people who live near elephant sanctuaries.

In 2002, *National Geographic* reported on pilot studies exploring the use of llama dung for filtering water. Researchers in Bolivia have discovered that llama dung contains microbes of the genus *Desulfovibrio,* which have the remarkable skill of neutralizing acidic water and aiding in the removal of dissolved metals, such as zinc, lead, copper, iron, and aluminum. The runoff from silver and tin mines is directed through ponds and lagoons filled with llama dung. Similar pilot studies have been done in the U.K., using cow and horse manure to filter runoff from old mines in Newcastle.

Despite all the research into uses for dung and the odd technological possibilities, I remain unconvinced that shit will become more than fuel and fertilizer. People in future centuries, however, may think me a silly old fool for having thought so un-creatively, with such low expectations of the innovativeness of my oh-so-clever species. Who knows: will we see shit-fueled computers made of bioplastic? Many things are possible, if only we put our minds to it. The biggest danger will be if we think we have found *the* solution, scale it up, and impose it on the world. If the study of complex social-ecological systems has taught us anything in the past few decades, it is that only through a mixture of public engagement, being aware of the many, diverse ramifications of any intervention in the world, and the creation of multiple, locally appropriate, adaptable resolutions will a sustainable and convivial human society be possible on this planet. Increasingly, "green" engineering companies, such as Enermodal in Canada, have recognized the mix of strategies, at several levels, that we need to address these complex issues.

Solutions for cities will combine variations of composting, energy generation, reuse of "gray" water (water that has been used for, say, having a shower) in toilets, reduced-water-flow toilets and showers, and conventional treatment plants. In agriculture, we are already seeing variations on the same theme.

The narratives we are co-constructing, in which we are all complicit, have many layers within them. The little girl I saw taking a dump in the communal refuse pile in Kathmandu lives in the same planetary family as the boy who dives into the outhouse in Mumbai so he can get a movie star's autograph in the film *Slumdog Millionaire*. This is the same world where I can find orchids on the toilets in the men's room of a Bangkok airport, and the same global village in which the people of Jakarta slums can watch their neighbor's partly digested dinner going by in open sewers. This world includes, as well, the cavernous cloacal system into which the good burghers of New York flush their poop, drugs, and whatever else might be passing through their intestines.

While some dream of giant bio-digesters, others are making high-tech vanilla, or electricity-producing toilets, or are composting chicken shit or human excrement for the garden. The good thing about this mixed-up timeline is that we can be sure that somewhere in the world, collectively as a species, we have all the ideas and technologies we need to solve the problem of excrement. Our bright and eco-friendly future may be in the hands and minds of a small child in Kolkata or a young engineer in California or a farmer in Germany. The natural tapestry into which our bodies have been woven may be frayed, moth-eaten, and faded, but the strands are still there, and although we might not be able to discern the meaning, there are still enough members of the original weavers' guild alive that we can re-create the vibrant colors of the original.

No matter how clever our technologies, they will only

be effective and helpful if they are designed for, and used in, the appropriate social-ecological contexts. The big question is not "Can we design new technologies based on shit?" (Of course we can.) The question is "Can we reorganize ourselves so that whatever technologies we devise promote a thriving, sustainable globe that is hospitable to our species?"

Can we devise a twenty-first-century notion of holonocratic citizenship that acknowledges our membership in an amazing, co-dependent, co-evolving panarchy of millions of plant, animal, and bacterial species, even as it recognizes the richness and inventiveness of human culture? Can we admit, freely and openly, that what we eat and how we handle our shit are essential acts of citizenship, as important as how we vote? I believe that the answers to all of these is: yes.

Farmers who anticipate future prices and climate restructure the landscape to arrive there, and in so doing change the future. If the anticipated future is based on monocultures and stability — that the future will be an extension of the past — farmers will suffer catastrophically. The upside of this malleability of the future — the fact that we can anticipate a certain kind of future and in so doing change it by what we do today — is that, with an understanding of shit, uncertainty, and complexity, we have a pretty good chance of enabling some pretty fine options for future generations.

I find, increasingly, that I am part of exciting discussions that go across scales, from households and farms to local and regional leaders, to representatives of global organizations, creating communities of practice that link across government, private businesses (large and small), and the general public.

We walk our lives along a fine line between self and

other, local and global, ecosystem conservation and ecosystem unfolding, tyranny and anarchy, hegemony and fragmentation. When I talk with my friends and colleagues in the local communities and communities of practice around the world, I really do feel as if I am part of an emerging holonocracy.

We can do this. No shit.

ACKNOWLEDGMENTS

For the writing of this book, I have drawn on my own work, that of friends and colleagues, and many hundreds of scholarly and not-so-scholarly articles. A list of many of these works is in the bibliography for those who wish to dig deeper. Portions of Chapter 8 were adapted from material I presented in the "Speakers of Renown" series of the Canada's International Development Research Centre in Ottawa on November 24, 2010, and material included in *Ecohealth: A Primer* by the author, ©2011 by Veterinarians without Borders/Vétérinaires sans Frontières – Canada, and appear by permission of the author. The latter document is freely downloadable from www.vwb-vsf.ca.

Thanks to Naomi Theodor for the book's title. The title was chosen to echo Darwin's *Origin of Species* (1872 edition); that it is also the title of a 1992 album by Brooklyn Gothic metal band Type O Negative, whom I had not heard of before, was a fortuitous discovery. Thanks to David Pearl, Eric Rumble, and others for chapter titles; to Kalisa Kwizera for his stories from Rwanda; to Andres Sanchez for the Mexico story; to Céline Surette for checking some of my calculations; and to Susilowati (Susi) Tana for the special gift of Kopi Luwak. Thanks to Crissy Boylan for her deft editing, and to Alexis Van Straten for helping me get a webpage launched.

Writing this book would have been impossible without a generous grant from the Canada Council for the Arts, the over-the-top diligence of my hard-working research assistant, Christina Grammenos, and the endless patience of my wife, Kathy.

DIGGING DEEPER: HOLD YOUR NOSE AND DIVE IN

Below is a brief selection of some of the references that I used in the writing of this book. For those who wish for more documentation: references, pictures, diagrams, and calculations are available at my website, DavidWaltnerToews.com. I also talked to a lot of people and drew on my own research and unpublished research of my colleagues around the world. Where appropriate, I have given that public acknowledgment. In some cases, they did not wish to have themselves or the communities where they worked identified.

Selected Bibliography

American Society of Agricultural Engineers (ASAE). 2003. "Manure production and characteristics" (http://www.manuremanagement.cornell.edu).

Barnes, D.S. 2005. "Confronting Sensory Crisis in the Great Stinks of London and Paris." In *Filth: Dirt, Disgust, and Modern Life*, ed. W.A. Cohen and

R. Johnson, 103–131. Minneapolis: University of Minnesota Press.

Bourke, J.G. 1891. *Scatalogic Rites of All Nations*. Washington: W.H. Lowdermilk & Company. [Catalog of cultural practices surrounding feces and urine.]

Brown, A.D. 2003. *Feed or Feedback: Agriculture, Population Dynamics, and the State of the Planet*. Utrecht: International Books. [Useful for an analysis of nutrient flows involved in current and historical agricultural practices.]

Chame, M. 2003. *Terrestrial Mammal Feces: A Morphometric Summary and Description*. Rio de Janeiro: Memórias do Instituto Oswaldo Cruz 98 (Suppl. I): 72–94 (http://www.scielo.br/scielo.php?pid=S0074-02762003000900014&script=sci_arttext&tlng=en). [Useful for the differences among animal scats.]

Claiborne, R. 1989. *The Roots of English: A Reader's Handbook of Word Origins*. New York: Doubleday.

Darimont, C.T., T.E. Reimchen, H.M. Bryan, and P.C. Paquet. 2008. "Faecal-Centric Approaches to Wildlife Ecology and Conservation: Methods, Data, and Ethics." *Wildlife Biology in Practice* 4 (2): 73–87.

Deutsch, L., and C. Folke. 2005. "Ecosystem Subsidies to Swedish Food Consumption from 1962–1994." *Ecosystems* 8(5): 512–528.

Flannery, T. 2007. *Tim Flannery: An Explorer's Notebook: Essays on Life, History, and Climate*. Toronto: HarperCollins Publishers Ltd., 202–203.

George, R. 2008. *The Big Necessity: Adventures in the World of Human Waste*. London: Portobello Books. [An excellent series of true stories about the new champions of human excrement management.]

Graves, R. 1955. *The Greek Myths. Volume 2.* Baltimore: Penguin Books.

Grove, R. 1976. "Coprolite Mining in Cambridgeshire." *Agricultural History Review* 24(1): 36–43.

Hanley, S.B. 1987. "Urban Sanitation in Preindustrial Japan." *Journal of Interdisciplinary History* 18(1): 1–26. [Shows the historical importance and value placed on night soil as a fertilizer in Japan.]

Hanski, I., and Y. Cambefort, eds. 1991. *Dung Beetle Ecology*. Princeton: Princeton University Press. [Comprehensive book on dung beetles.]

Jackson, W. 2003. "The Story of Civet." *The Pharmaceutical Journal* 271: 859–861.

Kauffman, S. 1995. *At Home in the Universe: The Search for the Laws of Self-organization and Complexity*. New York: Oxford University Press.

Key, N., and W. McBride. "The Changing Economics of U.S. Hog Production." Economic Research Report No. (ERR-52) 45 pp, December 2007 (http://www.ers.usda.gov /publications/err-economic-research-report/err52.aspx).

Lewin, R.A. 1999. *Merde: Excursions in Scientific, Cultural, and Socio-historical Coprology*. New York: Random House. [Excellent miscellany of information about the excrement of a variety of species.]

Lavery, T., R. Roudnew, P. Gill, J. Seymour, L. Seuront, G. Johnson, J.G. Mitchell, V. Smetacek. November 22, 2010. "Iron Defecation by Sperm Whales Stimulates Carbon Export in the Southern Ocean." *Proceedings of the Royal Society B: Biological Sciences*. 277 (1699): 3527–3353.

Madsen, M., B. Overgaard Nielsen, P. Holter, O.C. Pedersen, J. Brøchner Jespersen, K-M. Vagn Jensen, P.

Nansen, and J. Grønvold. 1990. "Treating Cattle with Ivermectin: Effects on the Fauna and Decomposition of Dung Pats." *Journal of Applied Ecology* 27(1): 1–15.

Mather, E., and J.F. Hart. 1956. "The Geography of Manure." *Land Economics* 32(1): 25–38. [Interesting article on global manure use.]

Nichols, E. S. Spector, J. Louzada, T. Larsen, T. Amezquita, M.E. Favila. 2008. "The Scarabaeinae Research Network: Ecological Functions and Ecosystem Services Provided by Scarabaeinae Dung Beetles." *Biological Conservation* 141(6): 1461–1474.

Ontario Ministry of Agriculture, Food, and Rural Affairs (OMAFRA). 2005. *Manure Management. Best Management Practices 16E* (https://www.publications.serviceontario.ca/ecom/MasterServlet/GetItemDetailsHandler?iN=BMP16E&qty=1&viewMode=3&loggedIN=false&JavaScript=y). [Note that OMAFRA has also published best management practices for "Nutrient Management Planning" and "Sewage Biosolids," both of which are relevant to this subject.]

Plain, R., J. Lawrence, and G. Grimes. 2012. "The Structure of the U.S. Pork Industry." *Pork Information Gateway* (www.porkgateway.org/PIGLibraryDetail/PF/1869.aspx#.UHRFiBgaCiY).

Platek, B. 2008. "Through a Glass Darkly: Miriam Greenspan on Moving from Grief to Gratitude." *The Sun Magazine* 385: 8.

Public Health Agency of Canada. 2000. "Waterborne Outbreak of Gastroenteritis Associated With a Contaminated Municipal Water Supply, Walkerton, Ontario, May–June 2000." *Canada Communicable*

Disease Report 26 (20) (http://www.phac-aspc.gc.ca /publicat/ccdr-rmtc/00vol26/dr2620eb.html).

Putman, R.J. 1983. "Carrion and Dung: Decomposition of Animal Wastes." *Studies in Biology* 156. London: Edward Arnold Publishers Limited. [Describes dung beetles and the process of decomposition.]

Reading, N.C., and D.L. Kasper. 2011. "The Starting Lineup: Key Microbial Players in Intestinal Immunity and Homeostasis." *Frontiers in Microbiology* 2: 148 (http://www.frontiersin.org/Cellular_and_Infection_Microbiology_-_closed_section/10.3389/fmicb.2011.00148/full)

Rockefeller, A. 1996. "Civilization & Sludge: Notes on the History of the Management of Human Excreta." *Current World Leaders* 39(6): 99–113. [Overview of the development of the sewage system.]

Sanderson, H., B. Laird, L. Pope, R. Brain, C. Wilson, D. Johnson, G. Bryning, A. Peregrine, A. Boxall, K. Solomon. 2007. "Assessment of the Environmental Fate and Effects of Ivermectin in Aquatic Mesocosms." *Aquatic Toxicology* 85: 229–240.

Scarabnet. http://www.scarabnet.org/ScarabNet /scarabnet_publications.html. [This network is a source of a wide range of publications and bibliographies on scarab research.]

Schoouw, N.L., S. Danteravanich, H. Mosbaek, J.C. Tiell. 2002. "Composition of Human Excreta: A Case Study from Southern Thailand." *The Science of the Total Environment* 286: 155–166.

Statistics Canada. 1996. "Estimated Livestock Manure Production. A Geographic Profile of Manure Production in Canada." Catalogue No. 16F0025XIB (http://www.statcan.ca/english/freepub/16F0025XIB/m/manure.pdf).

Steinfeld, H., P. Gerber, T. Wassenaar, V. Castel, M. Rosales, and C. de Haan. 2006. *Livestock's Long Shadow: Environmental Issues and Options*. Rome: Livestock, Environment and Development (LEAD) and Food and Agriculture Organization of the United Nations (FAO). [Useful overview on livestock agriculture in the twenty-first century.]

Tarr, J.A. 1975. "From City to Farm: Urban Wastes and the American Farmer." *Agricultural History* 49(4): 598–612. [Outlines the use of human waste in agriculture in the U.S., including guiding state regulations.]

Tarr, J.A., J. McCurley III, F.C. McMichael, and T. Yosie. 1984. "Water and Wastes: A Retrospective Assessment of Wastewater Technology in the United States, 1800–1932." *Technology and Culture* 25(2): 226–263. [Comprehensive, detailed, well-sourced account of the development of the U.S. sewage system, including the rise of the profession of sanitary engineering.]

Waltner-Toews, D., J. Kay, and N. Lister. 2008. *The Ecosystem Approach: Complexity, Uncertainty, and Managing for Sustainability*. New York: Columbia University Press.

Weiss, M.R. 2006. "Defecation Behavior and Ecology of Insects." *Annual Review of Entomology* 51: 635–661. [Excellent article describing a range of insect behavior surrounding defecation, full of interesting examples.]

Wotton, R.S., and B. Malmqvist. 2001. "Feces in Aquatic Ecosystems." *BioScience* 51(7): 537–544. [Good article outlining various aspects of feces in water-based habitats.]

OTHER BOOKS BY DAVID WALTNER-TOEWS

Non-fiction (author or co-author)

Ecohealth: A Primer. Veterinarians without Borders/ Vétérinaires sans Frontières – Canada, 2011. Available at http://www.vwb-vsf.ca/english/index.shtml under Resources.

Food, Sex, and Salmonella: Why Our Food Is Making Us Sick. Greystone Books, 2008.

Integrated Assessment of Health and Sustainability of Agroecosytems. With T. & M. Gitau. Taylor and Francis/CRC Press, 2008.

The Chickens Fight Back: Pandemic Panics and Deadly Diseases That Jump from Animals to Humans. Vancouver: Greystone Books, 2007.

Ecosystem Sustainability and Health: A Practical Approach. Cambridge: Cambridge University Press, 2004.

Good for Your Animals, Good for You: How to Live and Work with Animals in Therapy and Activity Programs and Stay Healthy. With A. Ellis. Pet Trust Foundation, 1994.

Food, Sex and Salmonella: The Risks of Environmental Intimacy. Toronto: NC Press, 1992.

One Animal Among Many: Gaia, Goats, and Garlic. Toronto: NC Press, 1991.

Non-Fiction (editor and co-author)

One Health for One World. Veterinarians without Borders/Vétérinaires sans Frontières – Canada, 2010. Available at http://www.vwb-vsf.ca/english/index .shtml under Resources.

The Ecosystem Approach: Complexity, Uncertainty, and Managing for Sustainability. With J. Kay and N. Lister (eds.). Also co-author on six chapters in this book. New York: Columbia University Press, 2008.

Poetry Collections

The Complete Tante Tina: Mennonite Blues and Recipes. Kitchener: Pandora Press, 2004.
The Fat Lady Struck Dumb. London, Ontario: Brick Books, 2000.
The Impossible Uprooting. Toronto: McClelland & Stewart, 1995.
Endangered Species. Winnipeg: Turnstone Press, 1988.
Good Housekeeping. Winnipeg: Turnstone Press, 1983.
The Earth Is One Body. Winnipeg: Turnstone Press, 1979.
That Inescapable Animal. Goshen, Indiana: Pinchpenny Press, 1974.

Fiction

Fear of Landing. Scottsdale, Arizona: Poisoned Pen Press, 2007.
One Foot in Heaven. Regina: Coteau Books, 2005.